Advances in Intelligent Systems and Computing

Volume 1281

The series "Advances in Intelligent Systems and Computing" contains publications on theory, applications, and design methods of Intelligent Systems and Intelligent Computing. Virtually all disciplines such as engineering, natural sciences, computer and information science, ICT, economics, business, e-commerce, environment, healthcare, life science are covered. The list of topics spans all the areas of modern intelligent systems and computing such as: computational intelligence, soft computing including neural networks, fuzzy systems, evolutionary computing and the fusion of these paradigms, social intelligence, ambient intelligence, computational neuroscience, artificial life, virtual worlds and society, cognitive science and systems, Perception and Vision, DNA and immune based systems, self-organizing and adaptive systems, e-Learning and teaching, human-centered and human-centric computing, recommender systems, intelligent control, robotics and mechatronics including human-machine teaming, knowledge-based paradigms, learning paradigms, machine ethics, intelligent data analysis, knowledge management, intelligent agents, intelligent decision making and support, intelligent network security, trust management, interactive entertainment, Web intelligence and multimedia.

The publications within "Advances in Intelligent Systems and Computing" are primarily proceedings of important conferences, symposia and congresses. They cover significant recent developments in the field, both of a foundational and applicable character. An important characteristic feature of the series is the short publication time and world-wide distribution. This permits a rapid and broad dissemination of research results.

Indexed by SCOPUS, DBLP, EI Compendex, INSPEC, WTI Frankfurt eG, zbMATH, Japanese Science and Technology Agency (JST), SCImago.

All books published in the series are submitted for consideration in Web of Science.

More information about this series at http://www.springer.com/series/11156

Ján Paralič · Peter Sinčák ·
Pitoyo Hartono · Vladimír Mařík
Editors

Towards Digital Intelligence Society

A Knowledge-based Approach

Springer

Editors
Ján Paralič
Department of Cybernetics and AI
Technical University of Košice
Košice, Slovakia

Pitoyo Hartono
Department of Electrical
and Electronics Engineering
Chukyo University
Nagoya, Japan

Peter Sinčák
Department of Cybernetics and AI
Technical University of Košice
Košice, Slovakia

Vladimír Mařík
Czech Institute of Informatics, Robotics
and Cybernetics
Czech Technical University in Prague
Prague 6, Czech Republic

ISSN 2194-5357 ISSN 2194-5365 (electronic)
Advances in Intelligent Systems and Computing
ISBN 978-3-030-63871-9 ISBN 978-3-030-63872-6 (eBook)
https://doi.org/10.1007/978-3-030-63872-6

This Springer imprint is published by the registered company Springer Nature Switzerland AG
The registered company address is: Gewerbestrasse 11, 6330 Cham, Switzerland

Preface

Humanity is now going through difficult times to fight the Covid-19 pandemic. Simultaneously, in these difficult times of physical separation, we can also realize how much digital society technology helps us cope with many difficulties that bring us this time. This book aims to provide readers with up-to-date knowledge on how to make these technologies smarter. We focus on selected research challenges for intelligent digital society and state-of-the-art methods how to face them.

The book's subtitle suggests that a core concept that the reader can study from various points of view in particular chapters is the knowledge. The knowledge that can help us intelligently face different digital society challenges (Part I of this book); the knowledge extracted from available big data employing intelligent analysis techniques (Part II). For efficient processing and analysis of data, there is a strong need for smart data and information modeling techniques (Part III).

This book was created within the context of the project entitled Knowledge-based Approaches for Intelligent Analysis of Big Data supported by the Slovak Research and Development Agency as project no. APVV-16-0213.

The process of preparation of this book had four phases. In the first one, potential authors could submit a one-page abstract of the proposed chapter in line with the main topics described in the call for chapters proposals. We received altogether 15 extended abstracts, all of them fitting pretty well into the book's theme. We afterward invited the authors to prepare their full chapter manuscripts, resulting in 14 chapter manuscripts submissions. In the next phase, two to three reviewers carefully reviewed each of the submitted chapter manuscripts. Based on the reviews, we accepted nine chapters. In the last phase, authors of accepted chapters prepared the final versions of their manuscripts, carefully incorporating the reviewers' suggestions. We are proud to present to you the result of this strict and demanding book preparation process.

We divided the content of the book into three parts, each one consisting of three chapters:

I. Digital Intelligence: Selected Research Challenges
II. Intelligent Analysis of Big Data
III. Data and Information Modelling

Part I presents selected challenges toward digital intelligence society. As the first challenge, the authors K. Machova et al. describe various forms of antisocial behavior in the online environment (mostly on social media), especially the creation and spreading of false information and using abusive language. Moreover, they also present intelligent approaches on how to cope with this challenge. The second chapter in this part of the book, written by I. Cik et al., focuses on five research challenges in advanced computer vision, where convolutional neural networks can help cope with these challenges intelligently. The last challenge presented by P. Blažek et al. is the combination of human intelligence with the power of machines to facilitate workloads or enable the mobility of their users.

Part II focuses on intelligent data analysis, as one of the pillars for the digital intelligence society. P. Bednar et al. present in their chapter two main types of knowledge-based approaches to data analysis. One of them is semantic modeling of data analytics processes, which can efficiently cover the explicit form of background knowledge. The second one is typical for medical applications, where a principal amount of background knowledge tends to stay tacit. In such a situation, human-in-the-loop approach is a way how to perform data analysis intelligently. V. Rozinajova et al. focus on intelligent analysis of data streams, particularly on two frequent data mining tasks, prediction and optimization. They provide an excellent summarization of their research work's selected outcomes, verified in the power engineering area in solving tasks of smart grid optimization. D. Andrešič et al. dive deeper into astronomical time series data analysis. They present an original approach that includes artificial neural networks enhanced by an evolutionary algorithm to find and learn the best performing classification model for preprocessed light curves.

Part III picks up some important data and information modeling challenges and solutions. V. Jirkovsky et al. present one of the key concepts of Industry 4.0, the seamless integration of various information resources providing a ubiquitous knowledge management system. L. Antoni et al. describe the computational methods for the generation of attribute implications in formal concept analysis and illustrate the process of attribute exploration for special types of object-attributes tables. Moreover, they proposed an own test for measuring pupils' computational thinking in several consecutive stages and collected data from pupils' national testing in Slovakia. M. Kvet and K. Matiasko cope in their chapter with one particular problem of database optimization. They propose an extension of the join operation method if the foreign key index is not present.

We want to thank all the authors for their valuable contributions to this book and all the reviewers for careful assessment and detailed reviews provided to the authors. All of them helped make this book an exciting work about selected topics on knowledge-based approaches suitable for solving some of the challenges of digital intelligence society.

Ján Paralič
Peter Sinčák
Pitoyo Hartono
Vladimír Mařík

Organization

Program Chairs

Ján Paralič	Technical University of Koáice, Slovakia
Vladimír Mařík	Czech Technical University in Prague, Czech Republic
Pitoyo Hartono	Chukyo University, Japan
Peter Sinčák	Technical University of Košice, Slovakia

Program Committee

Gabriela Andrejková	Pavol Jozef Šafárik University in Košice, Slovakia
Ľubomír Antoni	Pavol Jozef Šafárik University in Košice, Slovakia
František Babič	Technical University of Košice, Slovakia
Peter Bednár	Technical University of Košice, Slovakia
Ivana Budinská	Slovak Academy of Sciences, Slovakia
Marek Bundzel	Technical University of Košice, Slovakia
Radek Burget	Brno University of Technology, Czech Republic
Ľuboš Buzna	University of Žilina, Slovakia
Dalibor Fiala	University of West Bohemia, Czech Republic
Pitoyo Hartono	Chukyo University, Japan
Pavol Jancura	Eindhoven University of Technology, Netherlands
Aleš Janota	University of Žilina, Slovakia
Petr Kadera	Czech Technical University in Prague, Czech Republic
Stanislav Krajč	Pavol Jozef Šafárik University in Košice, Slovakia

Lenka Lhotská	Czech Technical University in Prague, Czech Republic
Vladimír Mařík	Czech Technical University in Prague, Czech Republic
Karel Mls	University of Hradec Králové, Czech Republic
Petr Novák	Czech Technical University in Prague, Czech Republic
Marek Obitko	Rockwell Automation, Czech Republic
Ján Paralič	Technical University of Košice, Slovakia
Libor Přeučil	Czech Technical University in Prague, Czech Republic
Jan Rauch	University of Economics, Prague, Czech Republic
Viera Rozinajová	Slovak University of Technology in Bratislava, Slovakia
Martin Sarnovský	Technical University of Košice, Slovakia
Peter Sinčák	Technical University of Košice, Slovakia
Pavel Tichý	Rockwell Automation, s.r.o., Czech Republic
Ján Vaščák	Technical University of Košice, Slovakia
Hiroaki Wagatsuma	Kyushu Institute of Technology, Japan
Petr Šaloun	Palacky University Olomouc, Czech Republic
Jan Šedivý	Czech Technical University in Prague, Czech Republic

Contents

Data and Information Modelling

Digital Intelligence: Selected Research Challenges

Addressing False Information and Abusive Language in Digital Space Using Intelligent Approaches

Kristina Machova[1(✉)], Ivan Srba[2], Martin Sarnovský[1], Ján Paralič[1], Viera Maslej Kresnakova[1], Andrea Hrckova[3], Michal Kompan[2], Marian Simko[2], Radoslav Blaho[4], Daniela Chuda[2], Maria Bielikova[2], and Pavol Navrat[2]

[1] Faculty of Electrical Engineering and Informatics, Technical University of Košice, Letná 9, 042 00 Košice, Slovakia
{kristina.machova,martin.sarnovsky,jan.paralic, viera.maslej.kresnakova}@tuke.sk

[2] Kempelen Institute of Intelligent Technologies, Mlynské Nivy II.18890/5, 811 09 Bratislava, Slovakia
{ivan.srba,michal.kompan,marian.simko,daniela.chuda, maria.bielikova,pavol.navrat}@kinit.sk

[3] Faculty of Informatics and Information Technologies, Slovak University of Technology, Ilkovičova 2, 842 16 Bratislava, Slovakia
andrea.hrckova@stuba.sk

[4] Faculty of Arts, Department of Psychology, Comenius University, Gondova Ulica 2, 811 02 Bratislava, Slovakia
radoslav.blaho@uniba.sk

Abstract. While digital space is a place where users communicate increasingly, the recent threat of COVID-19 infection even more emphasised the necessity of effective and well-organised online environment. Therefore, it is nowadays, more whenever in the past, important to deal with various unhealthy phenomena, that prohibit effective communication and knowledge sharing in the digital space. Undesired user behaviour and user-generated content in the online environment (mostly on social media) can have various forms, probably, the most harmful is the creation and spreading of *false information* (e.g., fake news) and using *abusive language* (e.g. hate speech). While a notable amount of research effort has been already dedicated to reducing and mitigating the negative consequences of such phenomena, a number of additional challenges and open problems remain unsolved. In this book chapter, at first, we provide a summary of existing research works, challenges and open problems. Consequently, we introduce our research results addressing false information and abusive language. Our approaches are based on intelligent and knowledge-based methods, mainly machine learning, natural language processing, and semi-automatic approaches. Besides the detection of undesired *content*, we present also less studied approaches to identify *users*, who contribute or spread such content. Our approaches are also complemented by the methods that are specific for computational social science and humanities as hyperlink network analysis, conceptual analysis, and interviews. Finally, we

J. Paralič et al. (Eds.): DISA 2020, AISC 1281, pp. 3–32, 2021.
https://doi.org/10.1007/978-3-030-63872-6_1

present some of our applications for monitoring and mitigating such undesired content and behaviour.

Keywords: False information · Abusive language · Digital space · Text processing · Machine learning

1 Introduction

Internet is a place where users communicate and express their thoughts. While most users tend to accept social norms in digital communication, others engage in undesired behaviour (misbehaviour or even anti-social behaviour), which negatively affects the rest of the community and its goals. Such malign behaviour in the online environment (mostly on social media) can have various forms, probably, the most harmful is the creation and spreading of *false information* and using *abusive language*. False information, which can be either spread accidentally (i.e., *misinformation*) or deliberately to deceive (i.e., *disinformation*), can be further categorised into more specific types, such as *fake news, fake reviews, rumours*, or *hoaxes*. Similarly, we distinguish different types of abusive language, such as *hate speech, trolling, cyberbullying, profanity* and in some cases, also *sexting.*

The negative consequences of such undesired online content on society are not negligible, particularly when a new coronavirus causing COVID-19 is threatening humans that shift their communication to digital space. Since some forms of regulation are in demand, computer-science research in connection with information science and psychology focuses on the development of *characterisation, detection* and *mitigation* methods and solutions. Among these, the detection task is being studied the most.

In this book chapter, we will at first define the terms and categorisation of concepts related to false information and abusive language. We will summarise solutions, challenges, and open problems, that characterise the state-of-the-art research on the characterisation, detection, and mitigation tasks. Consequently, we will present our contribution to this area by describing various approaches addressing these challenges and open problems.

Our main contributions are as follows:

- We provide an overview of previous research efforts and identify challenges and open problems that point out to future directions in the research focused on characterisation, detection and mitigation of false information and abusive language.
- We describe our research results while addressing such challenges and open problems.
- Finally, we provide a multidisciplinary perspective – while taking computer science as the primary point of view, we supplement it with our results from information science and psychology.

2 False Information

2.1 Definitions and Categorisation

In this chapter, we are mostly dealing with two of the most widespread forms of false information: fake news and fake reviews.

The concept of *fake news* is a neologism, often used to refer to a fictive message or disinformation [27] that are intentionally and veritably false and could mislead readers [3]. In practice, fake news is, however, a biased and widely-used term, often describing any false stories spreading on social media. Many authors therefore limit the definition of fake news to false news articles in the media, e.g. [27, 66]. In our conceptual analysis [39], we tended to adopt the latter definition, as computer science research often uses another term for false or mixed information on social media – rumours. Zubiaga [90] utilise the term rumours to denote items of information that are unverified at the time of posting. This definition is too general and does not help to determine, which posts on social media can be considered fake news and which rumours. Traditionally, rumours are characterised by the propagation of unauthorised messages that are of universal interest and are disseminated diffusely in social networks [9].

Due to the ambiguous utilisation of the term fake news and their overlap with some definitions of rumours, we differentiated the terms according to the production and distribution [39]. Fake news was distinguished as disinformation that was deliberately produced on the media by its author and rumours as misinformation that were distributed by users via social media. This is very important to discern as producers and distributors of false information have different intents. Producers create fake news either for profit, power or irresponsible reporting. Distributors (sometimes called also "useful idiots" [75]) are usually unaware of sharing false information. Their intents vary from the lack of knowledge or ignorance to the quest to socialise.

The fake news can be created and distributed on traditional media, partisan media or social networking sites. According to Fletcher [24], partisan media have a significantly lower number of readers and time spent on the web than traditional media. The impact of fake news, however, increased significantly with social networking sites. Any social network account can create professionally looking posts, which are spread quickly and for free [80]. Fake news distributed by Facebook have five times higher interaction than news from traditional media [50]. Spreading is affected by influencers much more than the quality or truth of the information.

Many users are ravished by emotions when sharing, as fake news often involve sensational headlines touching the basic emotions like fear, anger or surprise. If a misleading information comes from multiple sources in a similar period, it is not difficult to believe that it is serious information. The lack of information literacy also contributes to the quick spread of false information. Most social networks users do not dedicate enough time and space to confront the source of information and give them only a quick look. Another reason why users do not verify the truth of information is that the information is already enjoyed by thousands of users, and their sharing is perceived as some kind of recommendation or filter.

The producers of *fake reviews* have similar goals and methods as the producers of fake news, although the definitions and scope of news and reviews are different. Fake reviews can be defined as opinions on products or services that have been made to look valuable or genuine, often to deceive or mislead users. Lappas [48] perceives writing fake reviews as a form of attack, performed to purposefully harm or boost the reputation of an item (company, product, or person). The intents of fake news and reviews producers are similar – to make a profit, whereas in the case of fake news the intent is mixed with a need for power and/or ignorance [39]. The goals of both are to influence the opinions of recipients, and this could be a reason why these unethical activities are also called *opinion spamming*. Nevertheless, the term opinion spamming is mostly used for fake reviewers.

The repetition of lies or one-sided truths is called *propaganda* in politics, and the same applies for propagation in marketing. Both fake news and fake reviews are far beyond the ethical border. In both fields, tools like forged documentaries, planted evidence, staged media spectacles, and PR articles/content farms can be spotted [38]. Besides that, Sorgatz [75] designates other methods for describing discussion manipulation in the communities used in unethical marketing as well as for shifting public opinion in politics. Flooding the discussions with numerous comments of fake accounts to spread mass confusion and create the illusion of widespread support of a product or political party is called *astroturfing* and can be spotted in some campaigns. These fake accounts - *sock puppets* promote their ideology, create fake reviews and involve in the discussions. We classify both astroturfing and sockpuppeting in the anti-social behaviour, explained further in the Sect. 3.1.

2.2 State-of-the-Art Solutions, Challenges, and Open Problems

Fake News. The necessity for detection of false information in digital space significantly increased in the last years. We can witness growth in many research papers focused on (semi-)automatic detection of fake news, that employ a wide scale of different approaches – mostly based on machine learning. At the highest level, these approaches can be categorised according to the input data, which are considered during the extraction.

Probably the largest body of research derives for the purpose of the detection a number of *content-based features*, mostly describing text, while multimedia is used in the lesser extent (such as a number of words, a number of references, sentiment of the text, or various text representations). These approaches rely on an assumption, that style of fake news is different from true news. However, this kind of approaches has some drawbacks – features of style do not reflect the real veracity of the presented content, the predictions are difficult to explain, and finally, producers of fake news can quite easily mimic characteristics of true news and thus deceive trained detection models. At the same time, this group of approaches are widely applicable on different types of fake news (even coming from various sources, that were not covered by the training data) and do not require any domain-specific knowledgebases.

Another group of approaches analyses *contextual features* (such as information about the article's source, author or attached discussion). In comparison with content-based features, such contextual data cannot be influenced by the false information producer so easily. Thus they may provide useful supplementary information (contextual features are usually used in combination with content-based ones).

Some approaches focus specifically on *spreading of fake news* (such as a speed of spreading on social networks, spreading paths). Such spreading information has been confirmed to be a valuable feature that distinguishes between true and fake news. At the same time, it is usually difficult to obtain such data as they are available mostly only on social networking sites and not publicly available.

Finally, some authors aim to detect fake news more precisely (not only according to style or spreading characteristics), and thus they apply some kind of *existing knowledge-bases* and perform some sort of fact-checking of articles' content against prior verified knowledge. Such approaches may provide a detection with a higher precision and also a better explainability, however, they are fundamentally dependent on the extent of available knowledgebases and, moreover, mapping articles' content to the existing knowledge is a challenging task.

From the second perspective, existing approaches can be distinguished by the type of machine learning algorithms employed for the purpose of detection. At first, there are some approaches based on *traditional machine learning models*. For example, the work [88] is devoted to research the realistic mechanism of the fake news propagation. They tracked large databases of fake news and truth news from two different cultures – Japan and China. Spreading of fake news was monitored at early stages of propagation, e.g. five hours after the first posting. They declared that the information propagation at early stages offers novel features for the early detection of fake news. The novel features were set on the base of identifying topological properties of social networks. Another interesting result is that there are real differences in the propagation of fake news and true news. Another traditional approach to fake news detection based on machine learning methods is a work presented in [29]. It uses a vector representation of text data, TF-IDF weighting scheme and PCFG (Probabilistic Context-Free Grammars). The following machine learning methods were tested for use in solving the problem of recognition of fake news: Support Vector Machines, Stochastic Gradient, Random Forests and Naive Bayes. The best model was learned by Stochastic Gradient (the highest Accuracy = 0.687), while the baseline was a Naive Bayes model (Accuracy = 0.679).

Besides traditional machine learning models, *deep learning methods* have also been used to detect fake news. Several studies are focused on training a fake news classifier which would be able to detect the fake news solely based on the content of the article using NLP [6, 30]. Textual features represent the information extracted from the actual content of the article, which mostly includes the actual text of the article and its title.

While some of the deep learning models rely solely on the text when classifying the fake news, several approaches combine the content with the other contextual features. Linguistic features can be enhanced with multimedia content contained in the article [82]. Zhang in [87] presents the deep diffusive network based on Recurrent Neural Networks (RNN) and Gated recurrent unit (GRU) complemented by regularisation techniques. Authors use information related to the subject and also the author of a news article. More papers suggest that incorporation of integrated linguistic, syntactic, and semantic information about the articles could be beneficial, e.g., in [68] authors also used the data about the source of the article and response the article gets. More recently, a combination of *textual entailment* was used to improve the deep learning models. Authors in [70] presented such a system based on a combination of statistical machine learning and deep learning, which gained superior performance when compared to original approaches. An interesting approach to tackle the problem of obtaining a correctly annotated data was presented in [63]. Authors used crowdsourcing to collect the human-labelled data from seven different domains, and then compared the performance of various deep learning models when using original and human-labelled data. When using the social networks to spread the fake news, very important is also to study, how they are being propagated through the network. Long-Short Term Memory (LSTM) model was presented in [85] to represent the path, how a message spread.

Fake Reviews. The second problem, fake reviews identification is usually based on the machine learning methods used for training of prediction models. For example, work [47] presents algorithm REV2 based on Bayes classifier. The approach uses the following independent metrics: the credibility of an author, reliability of evaluation and quality of a product. The best testing result was promising (Precision = 0.846). The REV2 was employed in the detection system Flipkarte [47].

Another interesting question connected with the fake reviews' recognition is following. Is it necessary to process whole text of the reviews, or is it sufficient to process just the headlines of these reviews? The results obtained in [54] confirm that the models learned from the whole body of texts of reviews are more precise than those learned only from headlines. Another result was that Random Forest models were the best when learned from bodies of reviews (Accuracy = 0.983) and Naive Bayes models were best on the headlines data (Accuracy = 0.812).

The work [35] is also focused on the detection of fake reviews using a text analysis approach based on n-gram models and machine learning techniques. It compares six different classification techniques, namely, K-Nearest Neighbours (KNN), Logistic Regression, linear Support Vector Machine (SVM), Decision Trees and Stochastic Gradient Descent. To reduce the size of the lexical profile of texts, two methods were used: TF and TF-IDF weightings. The highest accuracy was achieved using the SVM algorithm and the lowest using the KNN algorithm. The paper [19] solved the problem of fake reviews uncovering using Naive Bayes classifier and Random Forest. The experiments showed that the Random Forest model achieved better results than the Naive Bayes classifier.

Authors of the work [16] studied an effectivity of a performance of the classification methods for fake reviews filtering when they are used in real-world scenarios that require online learning. Their datasets were from various domains as a trip advising, a recommendation of hotels, or an evaluation of restaurants. They used Bernoulli and Multinomial Naïve Bayes, KNN, Decision Trees, Random Forests and SVM. The best results were achieved by the SVM model (F1 measure = 0.899).

Challenges and Open Problems. The detection of fake news is accompanied with many challenges and open issues. At first, while some studies already analysed the characteristics of fake news (and how they differ from true news), there is a plenty room for additional especially large-scale studies that will focus on describing and understanding their characteristics and thus providing valuable input for feature engineering process performed as a part of detection methods. Despite the first approaches based on deep learning, a huge potential of such approaches (already demonstrated on many different NLP tasks) is still not completely revealed. Also, most of the existing approaches focus on the detection of individual pieces of content, while the research on detection of users who are possible producers or susceptible to share fake news is lacking. Last but not least, the advance in fake news detection is hampered by the absence of large well-annotated benchmark datasets – the existing datasets are labelled manually by human experts and very small (and thus unsuitable for learning some advanced models, such as deep learning ones) or labelled by some kind of simplified heuristic (e.g., the veracity of articles is determined by the reliability of source), what provides us with a large dataset but the labels may contain a lot of noise. This challenge could be addressed by appropriate interdisciplinary cooperation with the professionals from social sciences and humanities. For example, we identified professionals from the library and information science as suitable experts for labelling fake news articles within the preparation of our medical misinformation dataset.

Similarly, the detection of fake reviews poses several challenges and open problems. In general, fake reviews detection is dominantly solved by traditional machine learning approaches, e.g. SVM. There is a lack of annotated datasets with reliable identification of fake reviews. Similarly, as in the case of fake news, this is partially addressed by incorporating heuristic information, e.g. considering marginal opinions to be fake. However, even such datasets are often insufficient in terms of size. Deep learning approaches necessitate much large text corpora to fully leverage the potential of neural network. As a result of lacking large benchmark datasets, only a few approaches to fake news detection employ deep learning. The potential of state-of-the-art techniques (e.g. contextualised language models, transfer learning) in similar tasks are yet to be explored.

In the following subsections, we will present how we addressed some of these challenges and open problems as a part of our previous or ongoing work.

2.3 Characterisation of Fake News by Their Hyperlink Network

As mentioned above, there is a considerable room for research aimed at the characterisation of fake news. Specifically, we recognized an underestimation of the role of hyperlinks in fake news detection. Paradoxically, the quality of the links (resources) traditionally serves: 1) as indicators of the reliability of the research and also 2) as the input to website spam detection employed by some search engines. One of the mechanisms of the spam detection - Trust Rank is based on the premise that good websites seldom link to bad websites, and bad websites often link to good websites in an effort to increase the score in the hub. Trust Rank operates on the principle of selection of a small set of seed pages to be evaluated by an expert. When the reliable seed pages are manually identified, a crawl extending outward from the seed set seeks out similarly reliable and trustworthy pages [34]. The logic works oppositely as well, the closer a site is to spam resources, the more likely it is to be spam. Krishnan and Rashmi [46] call this opposite indicator Anti-Trust Rank.

The aim of our research [39] was to identify whether this linking principle is valid also for the fake news. A method of hyperlink network analysis was utilised to analyse the inbound and outbound hyperlinks of partisan news media (also labelled as fake news media by our local database). The dataset involved the hyperlinks of eighteen most popular mainstream news media and fifteen most popular partisan news media in Slovakia and Czech Republic. More than 171 million of domains of inbound and outbound hyperlinks of selected online news media were collected with Ahrefs tool[1], analysed and visualised with Gephi[2]. Ahrefs is a crawler, the original purpose of which is to

1 https://ahrefs.com/.
2 https://gephi.org/.

serve search marketing professionals. It collects, processes and stores data, consisting of hyperlinks, keywords and user behaviour. AhrefsBot is also mentioned as the second most active crawler on the Internet after Googlebot [86]. Gephi is an open-source software for visualisation and exploration of graphs and networks, applicable also for link analysis. Thanks to Gephi visualizations, the networks of directed hyperlinks from and to the online news media and the importance of some nodes, based on the automatic calculation of the sum of hyperlinks were obtained.

We concluded that there are some differences between the linking patterns of both types of media. The most significant one is that mainstream media rarely link to partisan media with dofollow attribute. There is also a little number of hyperlinks between mainstream media of different ownership, partisan media link to established mainstream media more often. We also identified some similarities between the links of mainstream and partisan news media. The vast majority of hyperlinks in both types of media is formed by directories or recommended articles from the same domain. Nevertheless, we confirmed the principle of Anti-Trust Rank and the role of hyperlinks for fake news detection. One of the possibilities for fake news detection could be manually and professionally selected blacklists of fake news media that would serve as seeds for other fake news media identification.

2.4 Detection of Fake News with Deep Learning

Deep learning has become one of the most popular approaches in machine learning in many fields during recent years. Especially in natural language processing (NLP) or computer vision, deep learning models have shown excellent results and proved to be very useful to tackle the problems from those domains.

In the past, many of the NLP problems were solved using traditional machine learning models. Many of these approaches faced issues related to very high-dimensionality of linguistic information. More recently, with the utilisation of more advanced representations such as word embeddings, neural networks gained popularity in NLP tasks thanks to the excellent results when compared to traditional machine learning models. Deep learning models have brought significant improvements over the n-grams model language. Such models can extract features on their hidden layers and learn the representations with high accuracy. These models can capture both semantic and syntactic patterns.

Many researchers have been using the popular deep learning methods such as Convolutional Neural Networks (CNN), Recurrent Neural Networks (RNN) or Long-Short Term Memory (LSTM). Especially RNN and LSTM models are popular when applying on the sequential data, as the model supports to memorise the results of previous computations and use that information in the current computation. This makes recurrent networks very well equipped to capture the context dependencies and makes them particularly useful in natural language processing because it helps to gain a deeper understanding of a language. Deep neural networks are also known for a fact that the larger the dataset is available for training, the better is the quality of extracted features. On the other hand, one of the significant drawbacks in certain types of such deep learning models could be the cost of computational resources needed to train these models. Training time increases dramatically, mostly due to the size of the datasets used to train such models.

Used Methods. *Convolutional neural networks* (CNN) are one of the deep neural networks which gained popularity in 2012, thanks to ImageNet [20]. However, they originated long before, in 1989, when they were first introduced to the world by computer scientist Yann LeCun in his work *Generalisation and network design strategies* [49]. CNNs can recognise and analyse not only images or video but have achieved success in NLP tasks or time sequence processing. During training, the individual layers are created as different degrees of abstraction of input data. Filter weights are initially initialised randomly. Gradually, the lower layers reveal low-level features, and the higher ones create more complex concepts. CNN topology is defined by three types of layers: a *convolutional layer*, a *pooling layer*, and a *fully connected layer*. A set of filters (kernel) represents the convolutional layer. The neuron's activation on the convolutional layer is calculated by moving the filter sequentially in the horizontal and vertical directions. The mathematical formula of the function:

$$S(i,j) = (K * I)(i,j) = \sum_m \sum_n I(m,n)K(i-m, j-n), \tag{1}$$

where K is the kernel, i.e., filter, and I is a digital image [31]. In the NLP tasks, usually one-dimensional convolutional neural networks (1D CNN) are used. An example of 1D CNN is depicted in Fig. 1.

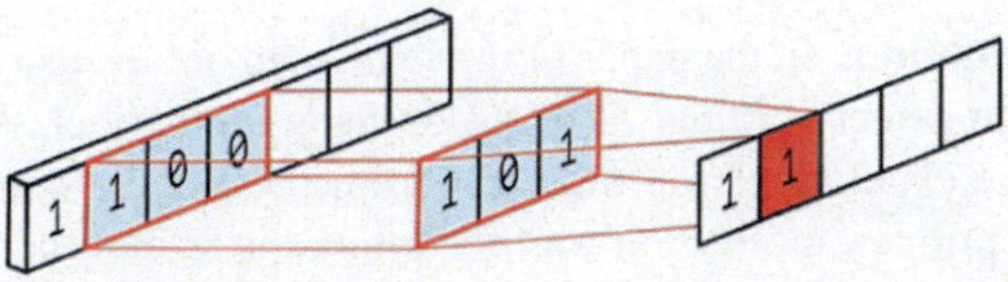

Fig. 1. One-dimensional convolutional process. Input data on the left, the filter in the middle, and the convolution output on the right.

Other types of deep neural networks are *recurrent neural networks* (RNN). RNNs are an excellent tool for NLP. Unlike standard neural networks, RNNs uses the word as an input, not the whole sample. RNN provides flexibility when working with the text of different lengths. RNN treats each word of the sentence as a separate input at time *t*, in addition to which it takes into account the activation value at time $t - 1$. A special type of RNN is *Long Short-Term Memory* (LSTM). LSTM has a complex structure of individual neurons and represents a way to deal with the vanishing gradient problem (see Fig. 2. The first figure shows the standard recurrent neural network and the vanishing gradient problem, which results in the loss of context. In contrast, the second figure shows the preservation of information and context in LSTM [32]). Using LSTM in any sequential task will ensure that long-term information and context is maintained.

The LSTM [37] network consists of memory blocks that are connected to layers instead of neurons. Each block contains gateways that manage the state and output of the block to regulate the information flow. Gateways can learn which data in a sequence is relevant and needs to be preserved. Thus, only relevant information will be included in the predictions.

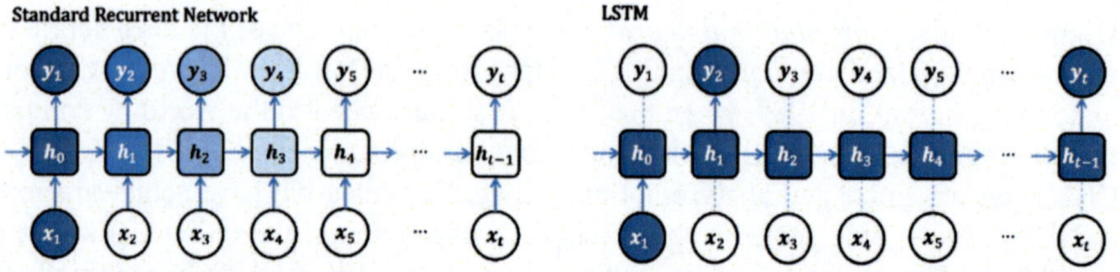

Fig. 2. The first figure shows the standard recurrent neural network and the vanishing gradient problem, which results in the loss of context. In contrast, the second figure shows the preservation of information and context in LSTM [32].

A special type of LSTM is *Bidirectional Recurrent Neural Network* (BiRNN), which can use context in both directions along the input sequence. BiRNNs consists of two separate hidden layers, one used for the positive time direction (forward states) while the other one for negative time direction (backward states). Both hidden layers are connected to the same output layer, allowing it to access each point's past and the future context in the sequence [72].

LSTM networks and their bidirectional variants BiLSTM are popular because they can learn how and when to forget certain information and also when not to use gateways in their architecture. The advantages of a one-way LSTM network are higher learning speed and better results.

Models and Evaluation. In the paper [45], we pointed out the use of CNN and LSTM neural networks to detect *political Fake News*, using a binary classification approach. We performed the experiments on a dataset from the Kaggle competition[3]. In these experiments, our primary motivation was to explore the size of the text needed to train such models. We used the entire full-texts of the news texts and compared the models trained using only the titles of political news articles. In both cases, we used the standard pre-processing of the text corpus: word tokenisation, the transformation of the text to lower case, removing punctuation, removing tokens that are not alphabetic, removing stop words. We used the word2vec implementation [55] to create word embeddings.

In the experiments, we compared the accuracy, precision, recall, and F1 score of the models. When solving fake news detection task, it is essential to train the models capable of recognising all fake news samples; therefore, we focused on the recall of the models. At the same time, as the training time of the neural network models can be significant, we wanted to determine whether the use of the architectures is also time-efficient; therefore, the training time was used as well.

The CNN used for text classification is depicted on Fig. 3. The input embeddings are connected to 1D convolutional layers with multiple filter widths and feature maps connected to the max-over-time pooling layer, which is connected to a fully connected layer with dropout and sigmoid function.

The LSTM model includes a layer of embeddings, LSTM (4), fully connected layer with 100 units and activation function ReLu, and fully connected layer with a sigmoid activation function. We used three regularisation layers – a dropout, and Adam optimisation [44]. The results of these experiments are shown in Table 1 and Table 2.

[3] Dataset available on https://www.kaggle.com/c/fake-news.

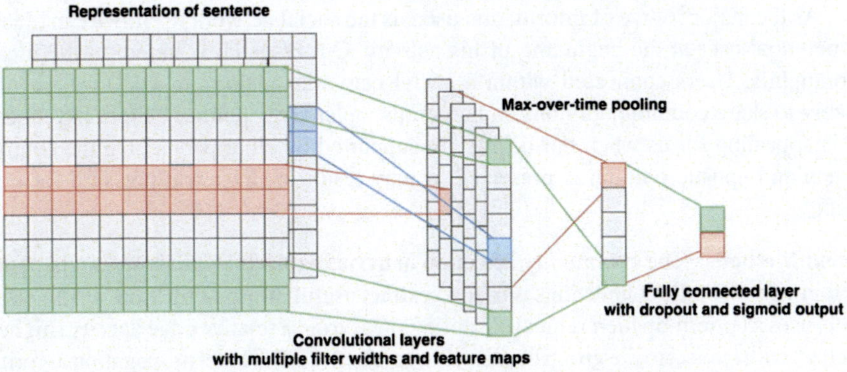

Fig. 3. Convolutional neural network for text classification.

Table 1. Results on the full-text data.

DL algorithm	Accuracy	Precision	Recall	F1 score	Time
CNN	**0.975**	0.969	**0.981**	0.975	82 s
LSTM	0.919	0.908	0.932	0.920	906 s

Table 2. Results on the title texts.

DL algorithm	Accuracy	Precision	Recall	F1 score	Time
CNN	**0.934**	0.909	0.981	0.933	3 s
LSTM	0.913	0.847	**1.000**	0.917	25 s

The results proved that the models trained on full-text data achieved better results as the models trained using only title text. However, the training of models on short texts could be significant in real-world tasks, when using much larger training data or when the deployed models must be updated frequently to adjust to incoming data.

2.5 Detection of Users Susceptible to Spread Fake News

The detection of fake news has been extensively researched in many domains. Usually, researchers are interested in a classification task of the news or their sources (in the mean of the publisher's media). In our research, we focus on the *user*, who is spreading the fake news on social networks. In other words, we explore the characteristics of social connections and fake news spreading. We believe that by identification of the fake news "spreaders" we will be able to mitigate the problem of misinformation on the social networks.

As the major source of information used, is the social network itself, we can identify a potential user at the beginning of his activity. Our research is based on the idea of homophily. Users connected within a social network (analogy to real life) are more likely to share common opinions on numerous topics (e.g., politics). From psychology, it is appealing for us when our beliefs are supported by other people. On the contrary, when an opposite opinion is presented, it may result in stress and pressure for many people.

Used Methods. The community detection in the networks is a widely researched topic. Often a graph of the connections is used to extract useful information. Louvain algorithm [11] uses a bottom-up hierarchical clustering measuring a relative edge density (higher is better) with respect to the group density (lower is better). The label propagation algorithm iteratively investigates the label of the neighbours, and the voting is applied. Another example of community detection is the usage of triangles of connections. Despite the more advanced methods that have been proposed as neural networks, thanks to the use of graph principles, standard approaches are still competitive.

An essential characteristic of a user in a social network is a centrality. Some users are influencing others, and vice versa. One of the most obvious metrics is the degree of centrality, which counts the user's connections. A more sophisticated approach is the betweenness centrality which uses the shortest paths. As the identification of the network "hubs" is an interesting problem, many other approaches have been used, e.g., PageRank.

To address these issues, we proposed a user model considering user's social context. The proposed user model consists of the following attributes:

- No. of followers
- No. of followees
- The ratio of followers and followees and vice versa
- The label of community a user belongs to
- Quantile of user followers and followings
- Degree of centrality
- Binary label if a user is a centroid
- No. of fake users followed by a user

Moreover, to these social context attributes, we utilised also the content of news, which user likes. For these, we store average values for:

- Text length
- No. of words in the title
- The average length of words in title and body
- No. of special symbols used in text
- No. of authors

This model is next used as an input to the standard classification task. In this way, we can identify users who are more likely to be fake news spreaders.

Models and Evaluation. To evaluate the proposed approach, we used the FakeNewsNet – BuzzFeed[4] dataset, which provides the user follower type of connections. Each user was labelled as a fake or a regular user, and the proposed user model was created. We found out that users tend to share only fake or true news. Moreover, we found out that the homophily effect is not as strong as we expected. This resulted in poor performance when only social context attributes were used (F1 = 0.56). On the contrary, when we also used the text attributes, the results were improved significantly (F1 = 0.99).

Based on the results, we can conclude that the social context of user can be used for fake user detection, but this is highly domain dependant. Each social network has a different characteristic, which can be utilised for fake news detection improvement.

2.6 Detection of Fake Reviews

Used Methods. We have used the following three methods for building prediction models for the fake review detection: Polynomial Naive Bayes, Random Forest, and Support Vector Machine. These methods were selected on the base of related works. Other reasons were that the Naive Bayes is often used as a baseline for experiments; the Random Forest is usually precise because it represents ensamble learning where the final decision is created by voting among many decisions of decorrelated trees, and Support Vector Machine is very precise because it forms the widest margin between examples of different classes.

Naive Bayes is a probabilistic classifier based on Bayes' theorem and independence assumption between features. The algorithm was widely studied in the 1950s. For n observations – features $\mathrm{x}_1, \ldots \mathrm{x}_\mathrm{n}$ the conditional probability for any class y_j can be expressed as below.

$$P\big(y_j/x_1,\ldots,x_n\big) = \beta P\big(y_j\big)\prod_{i=1}^{n} P\big(x_i/y_j\big) \tag{2}$$

This model is called the Naive Bayes classifier. Naive Bayes is often applied as a baseline for text classifying; however, its performance is reported to be outperformed by Support Vector Machines (SVM) [60]. SVM separates the sample space into two or more classes with the widest margin possible. The technique is originally a linear classifier; however, it can relatively efficiently perform non-linear classification by using a kernel. Continuing to complete the solution creating the widest margin between samples, it was observed that only the nearest points to the separating street determine its width. They are called support vectors. The objective is to maximise the width of the street, which is known as the primary problem of Support Vector Machines [7].

Random Forest [13] represents ensamble learning. It builds a set of de-correlated decision trees. It averages results of the set of decision trees to the final decision about class. The classification result is determined by voting. To ensure the condition that individual tree models must be independent, the random forest technique uses a random selection of attributes for each tree generation. We have selected Random Forest for the following advantages. Since the decision trees in the forest are independent, it is

[4] https://www.kaggle.com/mdepak/fakenewsnet.

suitable and easier to generate them in parallel. The random selection of a training set for training of each decision tree enables to validate (to test) it on data, which was not used for training. It facilitates validation. This approach is fast and accurate - so it has been used very often in recent years.

Models and Evaluation. The data were obtained from Amazon[5]. The dataset was created by 1 689 188 reviews from 192 403 customers on 63 011 products. The data from "Office Products" were used for training (53 258 reviews). The data pre-processing contained following steps: removing unsuitable symbols and unnecessary gaps, conversion to lowercase and deletion of punctuation, removal of "stop words", lemmatisation, tokenisation, creation of the document term matrix (DTM) using "Bag of words" representation, TF-IDF weighting scheme and word embeddings (Word2vec).

The input of the training consisted of tests of posts from dataset described above. Using a confusion matrix, following indicators of binary classification were quantified: Accuracy, Precision, Recall and F1 rate. The Table 3. ML models efficiency on data from Amazon (1 689 188 reviews on 63 011 products). shows the results of the effectivity of models learned by the selected machine learning methods. From the tested methods, the model learned using SVM achieved the best results.

Table 3. ML models efficiency on data from Amazon (1 689 188 reviews on 63 011 products).

ML algorithm	Accuracy	Precision	Recall	F1 rate
Naïve Bayes	0.557	0.704	0.667	0.685
Random Forest	0.658	0.703	0.912	0.794
SVM	**0.696**	**0.714**	**0.965**	**0.821**

3 Abusive Language

3.1 Definitions and Categorisation

The progress in information communication technologies and internet connection accessibility enables new social environment with many social interactions which can be desirable, as well as undesirable, e.g. in the form of an abusive language. This may include *anti-social internet behaviour* which violates norms or values of society or a specific social group [52] and differs from civil behaviour in online communities [18]. However, many substantial inconsistencies and substitutions can be found in commonly used terminology.

[5] http://jmcauley.ucsd.edu/data/amazon/.

Abusive language in the form of anti-social behaviour is part of the national language and provide a regular source of confronting and disturbing stories [65]. There are hundreds of definitions of such behaviour; therefore, it is quite difficult to find the exact meaning of this phenomenon [15] and is even a more significant challenge to do so in the online environment. In social sciences, anti-social behaviour is characterised as non-conformity, disregard, or unwillingness to adhere to rules and obligations imposed by society or social organisations [69]. Bacon et al. [5] states, such behaviour may include criminal acts that violate specific laws, as well as behaviour which is not illegal but contradict the social values and norms. Similarly, Ma [53] considers internet anti-social behaviour to be a part of social behaviour and distinguishes its forms, such as carrying out illegal activities, bullying others, cheating others, or illegal gambling. In an online environment, such a behaviour can be encouraged by dissociative anonymity and other factors of toxic disinhibition [78]. Even though anti-social behaviour is often considered to be the opposite of prosocial, some authors conclude they are not necessarily two poles of one dimension, but rather two separate dimensions and people can show any combination of both [21, 79]. Spreading rumours or sexting can serve as examples of this combination. This information is spread during the prosocial behaviour, but at the same time, the content is potentially hurtful for the reputation of the victim.

In general, the term *cyberaggression* is used for a behaviour which is intentional and aimed at harming other or others through electronic devices and perceived as aversive by the victim e.g. [8, 71]. One of the most studied topics in this area remains *cyberbullying* as a particular form of anti-social behaviour, perpetrated by a group or an individual, using electronic devices, which is characterized by intentionality to hurt, repetitiveness, and an imbalance of power [21, 73]. It affects social image, dignity or reputation of the victim. Willard [84] distinguishes several forms of cyberbullying, such as denigration, masquerade or impersonation, outing, bluejacking, happy slapping etc.

The term cyberbullying is sometimes confused with *trolling* e.g. [22], but the targets, methods and intents of these cyber aggressors differ [10]. Trolling can be defined as deceptive, destructive behaviour that violates the standards of the online community in which the troll operates e.g. [14, 36]. The main difference between the two terms is that trolling is an act, not targeted to a specific victim [17]. Trolls usually pretend to be legitimate participants in the conversation in an effort to gain the trust of other members of the online community. Later, trolls deliberately engage in upsetting the members of the community for their own amusement by sending malicious messages, posting aggressive arguments and provocative comments in public debates. Trolling strategies differ from individual to coordinated and from short-term to strategical long term acts. In the literature, two specific forms of trolling are mentioned: 1) flaming as an impulsive, aggressive verbal explosion, leading to a violent and "fiery" exchange of views and 2) griefing, which is similar to "malicious" behaviour especially in online games, including mutual communication of individual players [81].

Whereas trolling and its forms are usually seen as deviant or unethical behaviour, *cyberhate* is already classified as cybercrime. Cyberhate is clearly defined as the spread of *hate speech* (e.g. racist and xenophobic) to divide individuals in online communities based on various topics, such as political and religious beliefs, ethnic origin, sexual orientation, etc. [42].

Whether unethical or criminal behaviour, any act of creating and spreading manipulative, false or malicious information with an intent to deceive or hurt users is considered as anti-social behaviour. Anti-social behaviour as an act is challenging to detect or mitigate with computer science methods; therefore, we rather focus on the term abusive language as an analysable unit of this kind of behaviour.

3.2 State-of-the-Art Solutions, Challenges, and Open Problems

In contrast to false information detection tasks, where the main objective is to typically distinguish between two classes of textual content using binary classification approaches (e.g., real vs. fake news), abusive language detection usually involves the identification of different toxic or abusive language types. From the data analytical point of view, the problem can be considered as a multi-class classification task. In this case, the problem of unbalanced data is a common issue, as the frequency of occurrence of the different toxicity types may vary. Social networks present an essential source of conversational user data and can be used to train the models able to detect different types of abusive language appearing in discussions.

Besides traditional machine learning models, deep learning models are commonly used to identify the abusive language. Multiple topologies of CNN models were explored to find the most suitable model when handling with cyberbullying present in Tweets [2, 4]. Besides tweets, other data sources can be utilised to train the cyberbullying detectors. Authors in [1] used transfer learning within different datasets of conversational data (e.g. Wikipedia, Twitter) and then compared the performance of deep learning models. Multiple deep networks were successfully used in hate speech detection as well [64], including deep learning ensembles [89]. Also, in this case, data augmentation proved to be useful when complementing the deep learning model [67]

When monitoring social networks, an interesting aspect would be tracking the temporal aspects of toxicity in the comments. In [28], CNN model is proposed to detect the toxic tweets. Authors use hashtags related to toxic tweets and are also able to monitor the toxicity trend over time. Several previous works approached toxicity detection as the binary classification problem. But deep learning models are also used in more complex, ensemble approaches. In [40] an ensemble model consisting of CNN, Bi-LSTM and GRU is presented, which determines whether the text is toxic or not in the first step and then classifies the toxic comments into a more specific category representing the particular type of the toxicity.

One of the main challenges related to abusive language [77] is how to propose models that can not only classify benign and abusive language, but also perform a reliable distinction between various types of abusive language (e.g., hate speech vs. trolling). Also, an abusive language is very dynamic and can change significantly in time – the most of existing approaches cannot address such concept drifts sufficiently. Similarly, as in the case of false information, also the field of abusive language suffers from the lack of suitable large annotated datasets. There are several reasons for their absence, including copyright issues and difficult and subjective annotation process.

Remaining open problems [77] include multi-aspect detection (e.g., classification of hate speech target groups), multi-lingual and low-resource detection (e.g., for minor languages), detection utilizing contextual information or interpretable and explainable detection models.

3.3 Detection of Abusive Language with Deep Learning

In this section, we describe the use of deep neural networks models for the multi-label multi-class classification of comments toxicity in user conversational data. We mostly focused on the pre-processing of the textual content of the comments and the different ways of creating word embeddings.

Models and Evaluation. The dataset consists of comments from Wikipedia - an internet encyclopaedia with open content. This dataset was published as part of the Kaggle Competition - *Toxic Comment Classification Challenge*[6], which aimed to classify comments into the classes representing a specific type of toxicity: *toxic, severe toxic, obscene, thread, insult,* and *identity hate.*

Usually, the user-created conversational content, such as comments, often contains non-standard text, e.g. certain abbreviations, slang expressions, incomplete words, typos, etc. Standard text pre-processing methods could remove most of such content from the text. In this case (when solving such a specific type of task), the removal of this information can lead to the loss of essential traits crucial in the process of extracting features. Therefore, we decided to compare the deep learning models in terms of whether we use the standard pre-processing in them or not. Article [56] points to the same issue.

We performed experiments in which the text was classically pre-processed, as mentioned in the detection of fake news (Sect. 2.4), and in the second phase, we created only word tokens. We also compared word2vec [55], GloVe [62], and FastText [12] word embeddings. Each of the pre-trained word vectors has dimension of 300. The architecture consisted of a bi-directional LSTM network, a convolution layer, a sum of *global max pooling*, and a *global average pooling layer*, followed by a feedforward neural network that classifies examples into six classes. The results are shown in Table 4.

Table 4. Summary of the best results of neural networks in the detection of anti-social behaviour using multi-label classification. Marking 1, e.g. GloVe1 represents a model without pre-processing, labelled 2 with standard pre-processing. The training time was the same for all models - 4167 s.

	Accuracy	Loss	Precision	Recall	F1 score
GloVe 1	0.983	0.044	0.80	0.72	0.76
GloVe 2	0.982	0.047	0.79	0.71	0.75
Word2vec 1	0.983	0.052	0.82	0.68	0.74
Word2vec 2	0.982	0.049	0.82	0.65	0.73
FastText 1	0.983	0.045	0.82	0.70	0.76
FastText 2	0.982	0.052	0.83	0.65	0.73

[6] Dataset available on www.kaggle.com/c/jigsaw-toxic-comment-classification-challenge.

Models trained without the use of standard pre-processing were, on average, better than models trained with pre-processed texts. We confirmed the claims from the study [56] that in the fields such as abusive language detection in social media discussions, it is not appropriate to perform standard text pre-processing. Pre-processed data thus lose their individuality and specificity; therefore, models may achieve worse performance. It is clear from the experiments that the use of different embedding does not have a significant effect, although based on recall metric, GloVe1 seems to be the best option.

4 Application for Web Content Credibility Assessment

With the growing number of websites, the amount and variety of information available to the end-users increase enormously. The existence of inappropriate, misleading, or even deceptive web content may have severe consequences for people who increasingly rely on the Internet as the source of information on topics such as health, politics, or economy. In this section, we present our work focused on the automatic information extraction of properties affecting the credibility of the web articles. Firstly, we identify and extract several properties focusing on the article content (content-based), webpage structure (web-based), and rank (social-based) of the web site, where the article is published. We designed and implemented a web application that measures selected properties of texts published on the Internet. We have created a web browser extension that communicates with our application so the article can be evaluated while browsing the web. We evaluated our system on a small data sample of articles on selected topics from sources with different credibility.

4.1 Related Work

Much of previous credibility research has focused on understanding the factors that influence web content credibility ratings. This is not surprising, given that understanding the concept of "credibility" is not so easy and has many possible interpretations among researchers. Fogg conducted extensive studies to determine what factors people use to assess credibility [25, 26]. In practice, it was found that the appearance and first impression of the website had the most significant impact on the users' evaluation of content credibility.

[25] lists as many as 26 factors that affect credibility judgments. These factors were the result of the MAIN (Modality, Agency, Interactivity, Navigability) model. However, the model itself was the result of a theoretical analysis rather than empirical studies, like in [43] or [58]. With that number of factors listed in the MAIN model, the difficulty in selecting those relevant to the assessment of credibility also increases.

One of the most important areas where the credibility of web content is crucial is medicine. There are non-profit organisations, such as the Health on Net Foundation[7] or Quackwatch[8], that evaluate the reliability of medical websites. They do this by manually browsing the page to see if it meets certain conditions (such as quoting references, signing

[7] https://www.hon.ch/en/.
[8] https://quackwatch.org/.

articles by experts, etc.). However, this task is very challenging, not to mention the rapid growth of medical data on the web. Therefore, great efforts are being made to develop a system that would predict the credibility of web content without human intervention - automatically. [74] used controlled learning methods to predict the reliability of health-related websites.

[58] first identified a set of attributes affecting the trustworthiness of the web content and then used these attributes to predict the trustworthiness of new sites. They tested several machine learning algorithms (SVM, decision trees and Naive Bayes) where they initially considered up to 37 properties, but afterwards, they selected a subset of 22 of them achieving a higher accuracy of the classification task. These metrics were divided into 2 main categories, content-based, which were extracted from the page text and design, as well as social-based, which reflect the popularity of the website and the structure of the web links.

4.2 Methodology and Features Extraction

In our work, we focused primarily on the credibility of common websites, especially pages that publish news in the form of articles. Our goal was to find important features that affect web content credibility with the possibility to measure them automatically and visualise them in a suitable and useful way to the reader directly in his/her web browser when reading the article.

Based on previous research summed up above, we opted for metrics that come from three major categories – content-based, web-based, and social-based. Because each of the categories focuses on a different aspect, we gather more heterogeneous information that can later help us analyse the text (see Table 5. Web article properties (features) measured by our application).

Content-Based Features. We extract some easy to get basic properties from the content of the article, e.g. the number of words, the number of sentences, etc.). We also find out the number of personal sentences (i.e. sentences that contain characters like: ?, !, “”), difficult words (i.e. words with 4 or more syllables) and look for the occurrence of typical conspirator labels (like illuminates, chemtrails etc.) in the text. According to the analysis of manipulative techniques on selected Czech portals [33], labelling occurred in 18% of cases.

Similarly, as [58], we also perform sentiment analysis at the document level. The aim is to decide whether the article has an overall positive, negative or neutral sentiment.

One of the factors in assessing the quality of a text is its readability. This text property is defined as the reader's ability to understand a given text [51]. We measure this property via several readability indexes[9] (the last group of content-based features in Table 5. Web article properties (features) measured by our application).

Web-Based Features. Information about web links can be a suitable identifier that signals a lot about the credibility of a website [58]. For example, reliable pages often have a more complex structure and contain more links (internal and external) than simple pages, which can be created in minutes with untrusted content. The credibility of a given

9 https://readabilityformulas.com/.

Table 5. Web article properties (features) measured by our application.

Category	Features
Content-based	Number of characters
	Number of words
	Number of sentences
	Number of personal sentences
	Number of difficult words
	Labelling
	Article sentiment polarity
	The Flesch Reading Ease
	Simple Measure of Gobbledygook
	Coleman-Liau Index
	Automated Readability Index
	Flesch–Kincaid grade level
	Linsear Write Formula
	Gunning's Fog index
Web-structure	Number of all web links
	Number of internal web links
	Number of External web links
	Broken links
	Alexa Sites Linking in
	Contact link provided
	Privacy link provided
	Domain
Social-based	Alexa Rank
	Alexa Rank in Slovakia
	Google PageRank
	Conspiracy index based on konspiratori.sk

website is also affected by the presence of a privacy link (i.e. the link to a web page describing privacy policy of given website) and a contact link [74]. Previous studies have also shown [25, 26] that websites that are often unavailable or have many broken links are considered less trusted by users.

Another metric, suggesting a possibly credible website is the number of incoming links. Since the primary goal of the link should be directing users to the best possible information, it is presumed that a website will not link to sites that are less popular or untrustworthy. Instead, its purpose is to direct them to popular websites - because a popular website usually is the one, which has a good reputation and can be trusted [57].

The number of links is one of the most important factors when optimising a website for search engines. Therefore, we need to carefully consider this metric as decisive, because it is sometimes misused by search engine optimisation experts, utilising various forms of link building and non-ethical (black hat) SEO methods as, e.g. buying links for business purposes.

The context and quality of the incoming links together with their number is a more reliable indicator of website credibility. This is also a principle of Trust Rank and Anti-Trust Rank, mentioned above. However, the score of TrustRank or Anti-TrustRank is publicly unavailable., What is known is that TrustRank's reliability of a website diminishes with increased distance between the site and the seed set.

Social-Based Features From the category of social-based metrics, we focused on observing the rankings of website popularity worldwide, but also in Slovakia. A website's popularity can be a good indicator of its credibility [58]. Alexa rank expresses the degree of popularity of a particular website. With the help of the Alexa service, we can determine not only the global rank of a given website but also the rank in individual countries (in our case, e.g. we analysed articles from Slovak web sites only).

Another good indicator of web page credibility is PageRank [58, 74]. This metric provides information about the relative importance of a website and has been used successfully to improve web search performance. It uses the structure of web links to calculate the quality rating (PageRank) of individual web pages [59].

Due to the fact that we focus directly on Slovak web content, we also find out whether the given web page is in a database with not serious, misleading or propaganda content. This information is contained in a public database[10], in which the members of the commission evaluate all web sites suggested by users for inspection. The result of this initiative is an index of the site, which will be the average of the votes of the members of the commission who took part in the evaluation. The index ranges from 0 to 10. Web sites with an index of 6 or more are considered as non-credible pages.

4.3 Preliminary Results

Dataset. To perform this experiment, we created a dataset using our application. It ensured the analysis of individual articles by measuring selected features. The dataset consists of 54 articles published on the Internet with 27 measured properties. The articles come from 34 unique domains, where 31 are Slovak, and the remaining 3 are Czech. These internet media were selected based on a public database of websites from the *konspiratori.sk* initiative. Based on the data from this database, we determined whether it is credible or non-credible medium.

Published articles were selected based on the topic they discussed. We selected 4 topics for which we found the same number of credible as well as non-credible articles. Article topics (in parentheses is the number of articles):

- Etiology of Covid-19 virus, coronavirus as a weapon? (12);
- Migration, in particular events in relation to the Turkish border (16);

[10] https://www.konspiratori.sk/.

- Results of the 2020 parliamentary elections in Slovakia (14);
- Reports from the economic sphere (12).

Data-Preparation. Once we had the data collected, it was necessary to pre-process the data for further analysis. Three attributes contained missing values. We replaced the missing values in *broken_links* metric with the average value. For ranking metrics – *Alexa_rank, Alexa_rank_in_SVK*, we choose the maximum of the given column as the most suitable alternative to the NA values. As these are mainly untrustworthy websites, we believe that they are visited much less, so they are less popular. One of the next steps in the pre-processing was to derive the target attribute, *label*, from the *conspiracy_list* metric. The target attribute can have the following values:

- 0 – non-credible (*conspiracy list* value greater than 6);
- 1 – credible (*conspiracy list* value less than or equal to 6).

Data Understanding. We analysed the statistical characteristics of our data, including various visualisations. Especially useful were boxplot visualisations, where we looked for differences between the two target classes in the analysed features. The differences are interesting, especially for the web-based and social-based features.

The number of all links within the homepage but also the number of all backlinks to the given website is many times greater in the class of credible articles. On average, credible media has 3.22 times more links within a page (*num_all_links*), 3.23 times more backlinks (*Alexa_sites_linking_in*) and 3.66 times more links to another domain (*external_links*) than non-credible media. See the respective boxplots for two of these features in Fig. 4.

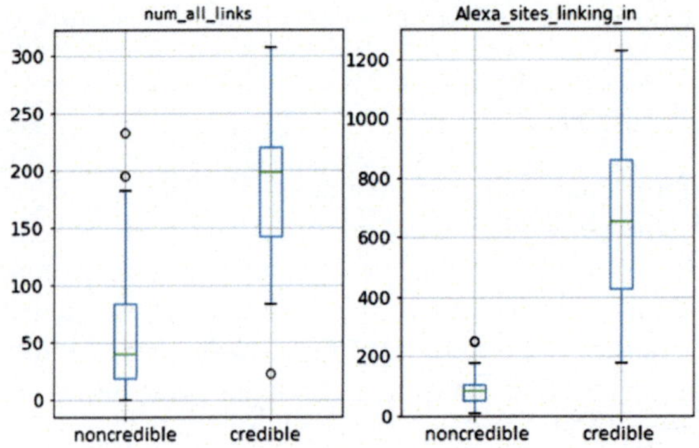

Fig. 4. Comparison of the number of all links and backlinks by the target value.

Social-based metrics also differed significantly in terms of the target attribute. The *Alexa_rank_in_SVK* metric had an average value of 55 (55th place in the given ranking) in the credible media, while the non-credible pages were around 294th place. The average placement of credible media in the Alexa rank was 45,861th place, while for the non-credible media it was up to 910,748th place (the scale of the y-axis in Fig. 5 is divided by 10^6). The average values of *Google_PageRank* were also higher for trusted media (5.9 - credible compared to 3.7 for non-credible).

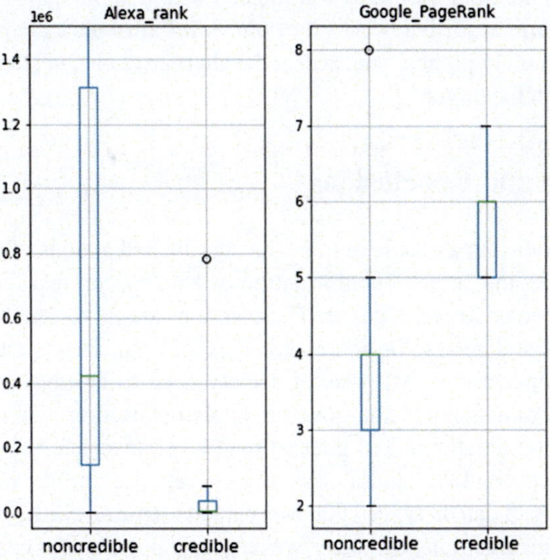

Fig. 5. Comparison of social properties to the target attribute.

Statistical Analysis. Aset of statistical analysis methods was used for evaluation. Graphs and descriptive characteristics indicated that there was a significant difference between the observed groups in web-based and social-based features with respect to the target classes of credible and non-credible web articles. Proper statistical tests have been used to verify findings from the data understanding phase statistically.

We used the Chi-square test for categorical attributes and the two-way Wilcoxon test for numeric attributes to test their dependence on the target attribute (with significance level alpha set on 0.05). We also looked at the relationship of attributes with the help of correlation analysis; the Kendall correlation coefficient was chosen.

Evaluation. We found out that there is a statistically significant difference between the averages of the two groups (credible and non-credible articles) in the *Google_PageRank, Alexa_sites_linking_in, Alexa_rank_in_SVK, Alexa_rank, num_all_links, internal_links, extrernal_links* and *broken_links* attributes. Based on the results of this test, we can state that these attributes depend on the target attribute.

Six attributes had an absolute value of the correlation coefficient (to the target) greater than 0.5. Positive correlation has been observed for *Google_PageRank* (0.71), *Alexa_sites_linking_in* (0.64), *num_all_links* (0.59), and *internal_links* (0.51). Negative correlation has been evidenced for *Alexa_rank_in_SVK* (-0.68) and *Alexa_rank* (-0.6).

We also looked at the relationship between selected attributes with each other. The *Google_PageRank* attribute is relatively strongly correlated with other attributes. A similar situation is with the second strongest attribute - *Alexa_rank_in_SVK*. Based on these results, we conclude that the *Google_PageRank* attribute has the most significant impact on the target attribute.

To verify the dependence of categorical attributes on the target, we used the Chi-square test. Four nominal attributes were tested with a significance level alpha set on 0.05. From the nominal attributes, only the *policy_link* attribute had a p-value = 0.029, so we can reject the null hypothesis and accept the alternative one, i.e. there is a dependence of this attribute on the target.

5 Discussion and Conclusions

Nowadays we are facing and able to trace the coordinated actions of people or bots that deliberately create and spread disinformation or hate speech that distorts trust, which is a basis for our coexistence in peace. These actions are motivated by political as well as financial goals and need to be addressed. This information war cannot be addressed by computer science alone. All parts of society need to be involved in this process, particularly: governmental and non-governmental organisations – to monitor the threats and take actions on a national and international level, teachers – to teach information literacy and critical thinking, and journalists – to report about the news in the highest level of objectivity. It can be argued that every good intention can be misused, e.g. by an authoritarian political regime, but today, we have a democracy that can be defended only with a coordinated action against disinformation and hate. Because it is disinformation and propaganda that are the weapons of authoritarian regimes.

Many legal institutions and governments recognised the severity of these threats (e.g., European Union) and prevention and elimination of false information and abusive language already became a responsibility of platform/website owners. The extensive amount of user-generated content, however, prohibit to rely only on manual force (human moderators). Therefore, there is a high demand for the proposal of various (semi-)automatic solutions that will assist or even completely replace manual detection that is a goal of computer science.

Such a goal is, nevertheless, still far to be achieved as current solutions (mostly based on supervised machine learning) can perform quite well, but their performance still provides plenty of room for improvement. In addition, labelling any information as false information or abusive language, either manually or (semi)-automatically, is somehow limited, simplistic, and relative. Philosophy has been questioning the existence of objective truth for decades. For example, Alfred North Whitehead [83] is famous for his statement: "*There are no whole truths. All truths are half-truths. It is trying to treat them as whole truths that plays the devil*". Also, a pragmatic philosopher, Feynmann [23] adds, that "*We never are definitely right, we can only be sure we are wrong.*" The aim of these statements is to warn us, that every today's knowledge does not have to be confirmed by tomorrow's experiment. Also, William James [41] notes that people construct their truths based on their thinking that is socially, culturally and historically biased and that we can only come closer to the real truth, but never will reach it ("*The true is only the expedient in our way of thinking, just as the 'right' is only the expedient in our way of behaving*"). The perception of the world is indeed subjective and, in this sense, we cannot talk about objective truth (and not on the Internet at all). Various opinions or rumours are common products of communication and as such, should not be in any way limited, as free speech is a cornerstone of democracy.

In this chapter, based on analyses of current research, we identified several open problems characterising computer-science approaches focused on fake news, fake reviews, hate speech and other types of abusive language. Consequently, we introduced some of our approaches addressing such open problems, which we proposed and evaluated as a part of our research projects dedicated to characterisation, detection, and mitigation of undesired users' behaviour in online space. In order to deploy the proposed methods and make their results available for end users, we proposed a browser plugin for automatic website credibility analysis. In addition, to reveal a high demand on appropriate datasets, we started to develop an universal and scalable platform Monant for monitoring and detection of false information and abusive language in the digital space [76]. Recently, we introduced an end-to-end system that uses a claim-based approach (claims being manually fact-checked by human experts), which utilizes information retrieval and machine learning techniques to detect medical misinformation [61].

In the future work, we plan to continue in the research of some additional more advanced approaches that will address some open problems that still remain unsolved – such a problem of the absence of large precisely annotated datasets. From this point of view, employment of active learning or semi-supervised learning seems promising. Finally focusing on some state-of-the-art solutions, such as deep learning or meta-learning, can result in other valuable contributions.

Acknowledgement. This work was partially supported by the Slovak Research and Development Agency under the contracts No. APVV-17-0267 and APVV SK-IL-RD-18-0004.

References

1. Agrawal, S., Awekar, A.: Deep learning for detecting cyberbullying across multiple social media platforms. In: Pasi, G., Piwowarski, B., Azzopardi, L., Hanbury, A. (eds.) Advances in Information Retrieval. ECIR 2018, pp. 141–153. Springer (2018). https://doi.org/10.1007/978-3-319-76941-7_11
2. Al-ajlan, M.A., Ykhlef, M.: Deep learning algorithm for cyberbullying detection. Int. J. Adv. Comput. Sci. Appl. **9**(9), 199–205 (2018). https://doi.org/10.14569/IJACSA.2018.090927
3. Allcott, H., Gentzkow, M.: Social media and fake news in the 2016 election. J. Econ. Perspect. **31**(2), 211–236 (2017). https://doi.org/10.1257/jep.31.2.211
4. Anindyati, L., Purwarianti, A., Nursanti, A.: Optimising deep learning for detection cyberbullying text in indonesian language. In: International Conference of Advanced Informatics: Concepts, Theory and Applications (ICAICTA), Yogyakarta, Indonesia, pp. 1–5 (2019). https://doi.org/10.1109/icaicta.2019.8904108
5. Bacon, A.M., Corr, P.J., Satchell, L.P.: A reinforcement sensitivity theory explanation of anti-social behaviour. Personality Individ. Differ. **123**(11), 87–93 (2018)
6. Bajaj, S.: The pope has a new baby! Fake news detection using deep learning. Technical report, Stanford University (2018)
7. Ben-Hur, A., et al.: Support vector clustering. J. Mach. Learn. Res. **2**(2), 125–137 (2001). https://doi.org/10.5555/944790.944807

8. Bauman, S., Underwood, M.K., Card, N.A.: Definitions: Another perspective and a proposal for beginning with cyberaggression. In: Bauman, S., Cross, D., Walker, J.L. (eds.), Principles of Cyberbullying Research: Definitions, Measures, and Methodology, pp. 41-45. Routledge, New York (2013)
9. Bergmann, J.R.: Discreet Indiscretions: The Social Organisation of Gossip. Aldine de Gruyter, New York (1993)
10. Blaho, R., Hrčková, A., Sabová, L., Mesárošová, B.: Anti-social behavior in online communities: terminology overview and hierarchy (2019). (Abstract retrieved from: https://www.icp2020.com/)
11. Blondel, V., Guillaume, J-L., Lambiotte, R., Lefebvre, E.: Fast unfolding of communities in large networks. J. Stat. Mech. Theory Exper, **10**, 1–12 (2008). IOP Publishing, https://doi.org/10.1088/1742-5468/2008/10/p10008
12. Bojanowski, P., et al.: Enriching word vectors with subword information. Trans. Assoc. Comput. Linguist. **5**, 135–146 (2017)
13. Breiman, L.: Random forests. Mach. Learn. **45**(1), 5–32 (2001). Kluwer Academic Publishers Hingham, MA, USA
14. Buckels, E.E., Trapnell, P.D., Paulhus, D.L.: Trolls just want to have fun. Personality Individ. Differ. **67**, 97–102 (2014)
15. Burney, E.: Making People Behave: Anti-Social Behaviour, Politics and Policy. Willan Publishing, Devon (2005)
16. Cardoso, E.F., Silva, R.M., Almeida, T.A.: Towards automatic filtering of fake reviews. Neurocomputing **309**, 106–116 (2018). https://doi.org/10.1016/j.neucom.2018.04.074
17. Cheng, J., Bernstein, M., Danescu-Niculescu-Mizil, C., Leskovec, J.: Anyone can become a troll: causes of trolling behavior in online discussions. In: Proceedings of the 2017 ACM Conference on Computer Supported Cooperative Work and Social Computing (CSCW 2017). Association for Computing Machinery, New York, NY, USA, pp. 1217–1230 (2017). https://doi.org/10.1145/2998181.2998213
18. Cheng, J., Danescu-Niculescu-Mizil, C., Leskovec, J.: Anti-social behavior in online discussion communities. In: ICWSM, AAAI Press, pp. 61–70 (2016)
19. Chowdhary, N.S., Pandit, A.A.: Fake review detection using classification. Int. J. Comput. Appl. **180**, 16–21 (2018)
20. Deng, J., et al.: Imagenet: a large-scale hierarchical image database. In: 2009 IEEE Conference on Computer Vision and Pattern Recognition, pp. 248–255 (2009)
21. Erreygers, S., Vandebosch, H., Vranjes, I., Baillien, E., De Witte, H.: Nice or naughty? The role of emotions and digital media use in explaining adolescents' online prosocial and anti-social behavior. Media Psychol. **20**(3), 374–400 (2017)
22. Ferenzi, N., Marshall, T.C., Bejanyan, K.: Are sex differences in anti-social and prosocial Facebook use explained by narcissism and relational self-construal? Comput. Hum. Behav. **77**(4), 25–31 (2017). https://doi.org/10.1016/j.chb.2017.08.033
23. Feynman, R.P.: The Character of Physical Law. Random House Publishing Droup, New York (1994). ISBN 13: 9780679601272
24. Fletcher, R., et al.: Measuring the reach of "fake news" and online disinformation in Europe. Reuters institute factsheet, (2018)
25. Fogg, B.J., et al.: What makes Web sites credible? a report on a large quantitative study. In: Proceedings of the SIGCHI Conference on Human Factors in Computing Systems, Seattle, pp. 61–68 (2001)
26. Fogg, B.J., Soohoo, C., Danielson, D.R., Marable, L., Stanford, J., Tauber, E.L.: How do users evaluate the credibility of Web sites? a study with over 2,500 participants. In: Proceedings of the 2003 conference on Designing for user experiences, San Francisco, pp. 1–15 (2003)
27. Gelfert, A.: Fake news: a definition. Informal Logic **38**(1), 84–117 (2018)

28. Georgakopoulos, S.V., Tasoulis, S.K., Vrahatis, A.G., Plagianakos, V.P.: Convolutional neural networks for twitter text toxicity analysis. Oneto, L., Navarin, N., Sperduti, A., Anguita, D. (eds) Recent Advances in Big Data and Deep Learning. INNSBDDL 2019. Proceedings of the International Neural Networks Society, vol. 1. Springer (2020)
29. Gilda, S.: Evaluating machine learning algorithms for fake news detection. In: Pune Institute of Computer Technologies, Pune, India, IEEE 15th Student Conference on Research and Development (SCOReD), pp. 110–115 (2017)
30. Girgis, S., Amer, E., Gadallah, M.: Deep learning algorithms for detecting fake news in online text. In: Proceedings - 2018 13th International Conference on Computer Engineering and Systems ICCES 2018, pp. 93–97 (2019)
31. Goodfellow, I., Bengio, Y., Courville, A.: Deep Learning, MIT Press (2016). ISBN: 0262035618
32. Graves, A.: Supervised sequence labelling. In: Supervised Sequence Labelling with Recurrent Neural Networks. Studies in Computational Intelligence, vol. 385, Springer, Berlin, pp. 5–13 (2012). https://doi.org/10.1007/978-3-642-24797-2_2
33. Gregor, M., Vejvodová, P.: Analysis of manipulative techniques on selected Czech servers. Department of International Relations and European Studies, Masaryk University, (2016)
34. Gyongyi, Z., Garcia-Molina, H., Pedersen, J.: Combating web spam with trustrank. In: Proceedings of the 30th International Conference on Very Large Data Bases, Volume 30 (VLDB 2004), VLDB Endowment, pp. 576–587 (2004)
35. Hadeer, A., Issa, T., Sherif, S.: Detection of online fake news using N-gram analysis and machine learning techniques, LNCS 10618, pp. 127–138. Springer (2017)
36. Hardaker, C.: "Uh. . . . not to be nitpicky,,,,,but…the past tense of drag is dragged, not drug.": An overview of trolling strategies. J. Lang. Aggression Conflict, **1**(1), 58–86 (2013)
37. Hochreiter, S., Schmidhuber, J.: Long short-term memory. Neural Comput. **9**(8), 1735–1780 (1997)
38. Holiday, R.: Trust Me, I'm Lying: Confessions of a Media Manipulator. Penguin, New York (2013). 320 p.
39. Hrckova, A., Srba, I, Moro, R., Blaho, R., Simko, J., Navrat, P, Bielikova, M.: Unravelling the basic concepts and intents of misbehavior in post-truth society. Bibliotecas. Anales de Investigacíón **15**(3) (2019)
40. Ibrahim, M., Torki, M., El-Makky, N.: Imbalanced toxic comments classification using data augmentation and deep learning. In: 2018 17th IEEE International Conference on Machine Learning and Applications (ICMLA), Orlando, FL, 2018, pp. 875–878 (2018). https://doi.org/10.1109/icmla.2018.00141
41. James, W.: The Meaning of Truth, A Sequel to Pragmatism (1909). ISBN 9781534647145
42. Kaakinen, M., Räsänen, P., Näsi, M., Minkkinen, J., Keipi, T., Oksanen, A.: Social capital and online hate production: a four country survey. Crime Law and Social Change **69**(1), 25–39 (2018)
43. Kakol, M., Nielek, R., Wierzbicki, A.: Understanding and predicting web content credibility using the content credibility corpus. Inf. Process. Manag. **53**, 1043–1061 (2017)
44. Kingma, D.P., Ba, J.: Adam: a method for stochastic optimisation. arXiv preprint arXiv:1412.6980 (2014)
45. Krešňáková, V.M., Sarnovský, M., Butka, P.: Deep learning methods for Fake News detection. In: 2019 IEEE 19th International Symposium on Computational Intelligence and Informatics and 7th IEEE International Conference on Recent Achievements in Mechatronics, Automation, Computer Sciences and Robotics (CINTI-MACRo), pp. 143–148. IEEE, (2019)
46. Krishnan, V., Rashmi R.: Web spam detection with anti-trust rank. In: AIRWeb, vol. 6 (2006)
47. Kumar, S., Kumar, M., Hooi, B., Faloutsos, CH., Makhija, D., Subrahmanian, V.S.: REV2: Fradulent user prediction in rating platforms. Stanford (2018). https://cs.stanford.edu/~srijan/pubs/rev2-wsdm18.pdf. Accessed 23 June 2020

48. Lappas, T.: Fake reviews: the malicious perspective. In: International Conference on Application of Natural Language to Information Systems. Springer, Heidelberg (2012)
49. LeCun, Y.: Generalisation and network design strategies. Connectionism Perspect. **19**, 143–155 (1989)
50. Lewandowsky, S., et al.: Beyond misinformation: understanding and coping with the "Post-Truth" era. J. Appl. Res. Memory Cogn. **6**(4), 353–369 (2017). https://doi.org/10.1016/j.jarmac.2017.07.008
51. Louis, A.: Predicting Text Quality: Metrics for Content, Organization and Reader Interest. University of Pennsylvania (2013)
52. Ma, H.K.: Internet addiction and antisocial internet behavior of adolescents. Sci. World J. **11**, 2187–2196 (2011)
53. Ma, H.K., Chu, M.K.Y., Chan, W.W.Y.: Construction of a teaching package on promoting prosocial internet use and preventing antisocial internet use. Sci. World J. **11**, 2136–2146 (2011)
54. Machova, K., Mach, M., Demkova, G.: Modelling of the fake posting recognition in online media using machine learning. In: SOFSEM 2020 - 46th International Conference on Current Trends in Theory and Practice of Computer Science, Limassol, Cyprus, pp. 1–9, Springer, Heidelberg, (2020)
55. Mikolov, T., et al.: Efficient estimation of word representations in vector space. arXiv preprint arXiv:1301.3781 (2013)
56. Mohammad, F.: Is preprocessing of text really worth your time for online comment classification?. arXiv preprint arXiv:1806.02908 (2018)
57. Monitor Backlinks. https://monitorbacklinks.com/blog/seo/why-backlinks-are-important. Accessed 30 Apr 2020
58. Olteanu, A., Peshterliev, S., Liu, X., Aberer, K.: Web credibility: features exploration and credibility prediction. In: Serdyukov, P., et al. (eds.) Advances in Information Retrieval. ECIR 2013. Lecture Notes in Computer Science, vol. 7814. Springer, Berlin, Heidelberg, (2013)
59. Page, L., Brin, S., Motwani, R., Winograd, T.: The PageRank Citation Ranking: Bringing Order to the Web. The Standford University (1999)
60. Pang, B., Lee, L.: Vaithyanathan, S.: Thumbs up? sentiment classification using machine learning techniques. In: Proceedings of the Conference on Empirical Methods in Natural Language Processing (EMNLP), Philadelphia, pp. 79–86 (2002)
61. Pecher, B., Srba, I., Moro, R., Tomlein, M., Bielikova, M.: FireAnt: claim-based medical misinformation detection and monitoring. In: Proceedings of European Conference on Machine Learning and Principles and Practice of Knowledge Discovery in Databases - ECML-PKDD 2020, Springer (2020, to appear)
62. Pennington, J., Socher, R., Manning, Ch.: GloVe: Global vectors for word representation. In: Proceedings of the 2014 Conference on Empirical Methods in Natural Language Processing (EMNLP), pp. 1532–1543 (2014)
63. Pérez-Rosas, V., Kleinberg, B., Lefevre, A., Mihalcea, R.: Automatic detection of fake news. In: Proceedings of the 27th International Conference on Computational Linguistics (COLING), Association for Computational Linguistics, pp. 3391–3401 (2018)
64. Ranasinghe, T., Zampieri, M., Hettiarachchi, H.: BRUMS at HASOC 2019: deep learning models for multilingual hate speech and offensive language identification. In: CEUR 2019 Workshop Proceedings, (2019)
65. Riley, D.: Anti-social behaviour: children, schools and parents. Education and the Law **19**(3–4), 221–236 (2007)
66. Rini, R: Fake news and partisan epistemology. Kennedy Institute Ethics J. **27**(2) (2017)

67. Rizos, G., Hemker, K., Schuller, B.: Augment to prevent: short-text data augmentation in deep learning for hate-speech classification. In: Proceedings of the 28th ACM International Conference on Information and Knowledge Management (CIKM 2019). Association for Computing Machinery, New York, NY, USA, pp. 991–1000 (2019). https://doi.org/10.1145/3357384.3358040, (2019)
68. Ruchansky, N, Seo, S, Liu, Y.: CSI: a hybrid deep model for fake news detection. In: Proceedings of the 2017 ACM on Conference on Information and Knowledge Management, pp. 797–806 (2017)
69. Rutter, M.: Commentary: causal processes leading to antisocial behaviour. Dev. Psychol. **39**, 372–378 (2003)
70. Saikh, T., Anand, A., Ekbal, A., Bhattacharyya, P.: A novel approach towards fake news detection: deep learning augmented with textual entailment features. In: Métais E., Meziane F., Vadera S., Sugumaran V., Saraee M. (eds) Natural Language Processing and Information Systems. NLDB 2019. Lecture Notes in Computer Science, vol. 11608. Springer (2019)
71. Schoffstall, C.L., Cohen, R.: Cyber aggression: the relation between online offenders and offline social competence. Soc. Dev. **20**(3), 587–604 (2011)
72. Schuster, M., Paliwal, K.K.: Bidirectional recurrent neural networks. IEEE Trans. Sig. Process. **45**(11), 2673–2681 (1997)
73. Smith, P.K., Mahdavi, J., Carvalho, M., Fisher, S., Russell, S., Tippett, N.: Cyberbullying: its nature and impact in secondary school pupils. J. Child Psychol. Psychiatry **49**, 376–385 (2008). https://doi.org/10.1111/j.1469-7610.2007.01846.x
74. Sondhi, P., Vydiswaran, V.G.V., Zhai, C.: Reliability prediction of Webpages in the medical domain. In: Baeza-Yates, R., et al. (eds.) Advances in Information Retrieval ECIR 2012. Lecture Notes in Computer Science, vol. 7224, pp. 219–231. Springer, Heidelberg (2012)
75. Sorgatz, R.: The encyclopedia of misinformation a compendium of imitations, spoofs, delusions, simulations, counterfeits, impostors, illusions, confabulations, skullduggery, frauds, pseudoscience, propaganda, hoaxes, flimflam, pranks, hornswoggle, conspiracies & miscellaneous fakery. Abrams, New York (2018)
76. Srba, I., Moro, R., Simko, J., et al.: Monant: universal and extensible platform for monitoring, detection and mitigation of antisocial behaviour. In: Proceedings of WS on Reducing Online Misinformation Exposure - ROME 2019. pp. 1–7 (2019)
77. Srba, I., Lenzini, G., Pikuliak, M., Pecar, S.: Addressing hate speech with data science: an overview from computer science perspective. In: Wachs, S., Koch-Priewe, B., Zick, A. (eds.) Wenn Hass spricht. Springer (2021, to appear)
78. Suler, J.: The online disinhibition effect. Cyberpsychol. Behav. **7**(3), 321–326 (2004)
79. Veenstra, R.: The development of Dr. Jekyll and Mr. Hyde: Prosocial and antisocial behavior in adolescence. In: Fetchenhauer, D., Flache, A., Buunk, A.P., Lindenberg, S. (eds.) Solidarity and Prosocial Behavior: An Integration of Sociological and Psychological Perspectives, pp. 93–108. Springer, Heidelberg (2006). https://doi.org/10.1007/0-387-28032-4_6
80. Vítek, F.: Fake news – where did it begin and where do we go? (2020). http://mocnedata.sk/2018-fake-news/. Accessed Jun 2020
81. Voggeser, B.J., Singh, R.K., Göritz, A.S.: Self-control in online discussions: disinhibited online behavior as a failure to recognise social cues. Front. Psychol. **8**(2372), 1–11 (2018). https://doi.org/10.3389/fpsyg.2017.02372
82. Wang, F., et al.: EANN: event adversarial neural networks for multi-modal fake news detection. In: Proceedings of the 24th ACM SIGKDD International Conference on Knowledge Discovery & Data Mining - KDD 2018, pp. 849–857 (2018). https://doi.org/10.1145/3219819.3219903
83. Whitehead, A.N.: Dialogues: Prologue (1954)
84. Willard, N.E.: Cyberbullying and cyberthreats: responding to the challenge of online social aggression, threats, and distress. Research Press (2007)

85. Wu, L., Huan, L.: Tracing fake-news footprints: characterizing social media messages by how they propagate. In: WSDM 2018 - Proceedings of the 11th ACM International Conference on Web Search and Data Mining, pp. 637–645. Association for Computing Machinery (2018). https://doi.org/10.1145/3159652.3159677
86. Zeifman, I., Breslaw, D.: A closer look at the most active good bots (2017). https://www.imperva.com/blog/most-active-good-bots/
87. Zhang, J., Cui, L., Fu, Y., Gouza, F.B.: Fake news detection with deep diffusive network model. ArXiv, abs/1805.08751 (2018)
88. Zhao, Z., Zhao, J., Sano, Y., Levy, O., Takayasu, H., Takayasu, M., Li, D., Wu, J., Havlin, S.: Fake news propagates differently from real news even at early stages of spreading. EPJ Data Sci. **9**(7), 1–14 (2020). https://doi.org/10.1140/epjds/s13688-020-00224-z
89. Zimmerman, S., Fox, C., Kruschwitz, U.: Improving hate speech detection with deep learning ensembles. In: LREC 2018 - 11th International Conference on Language Resources and Evaluation, pp. 2546–2553 (2019)
90. Zubiaga, A., Aker, A., Bontcheva, K., Liakata, M., Procter, R.: Detection and resolution of rumours in social media: a survey. ACM Comput. Surv. **51**(2) (2018). https://doi.org/10.1145/3161603

Application and Perspectives of Convolutional Neural Networks in Digital Intelligence

Ivan Čík(✉), Miroslav Jaščur, Andrinandrasana David Rasamoelina, Ján Magyar, Lukáš Hruška, Fouzia Adjailia, Marián Mach, Marek Bundzel, and Peter Sinčák

Department of Cybernetics and Artificial Intelligence, Technical University of Košice, Košice, Slovakia
{ivan.cik,miroslav.jascur,andrijdavid,jan.magyar,lukas.hruska, fouzia.adjailia,marian.mach,marek.bundzel,peter.sincak}@tuke.sk

Abstract. Convolutional neural networks are the state-of-the-art approach for advanced computer vision tasks, as they offer capabilities beyond straightforward application for image processing. This review provides an introduction to five areas where convolutional neural networks are a core topic of research: 2D and 3D object classification, image segmentation, few-shot learning, reinforcement learning and explainability of neural networks. Each section provides an introduction to the research topic, identifies the main research questions, and lists modern solutions to these problems.

Keywords: Convolutional neural networks · Deep learning · Digital intelligence · Explainable AI · Few-shot learning · Meta-learning · Object recognition · Reinforcement learning · Segmentation

1 Introduction

Methods of artificial intelligence, and machine learning in particular, have found their application in various fields of scientific research but also everyday applications. Humanity produces an unparalleled amount of data, and this trend will only speed up as we enter the era of big data, Internet of Things, and Industry 4.0. The immense success of machine learning models and applications is primarily due to the ability of deep learning (DL) methods to process data and transform it so it can be used for prediction tasks. Although simple feed-forward neural networks (FNNs) can process numerical data, they are unable to consider spatial information, which limits their use, since a large proportion of data is either visual, or can be represented as images. This results in poor support for spatial invariance, meaning that neural networks would perform poorly on inputs that are different from those in the training set.

J. Paralič et al. (Eds.): DISA 2020, AISC 1281, pp. 33–58, 2021.
https://doi.org/10.1007/978-3-030-63872-6_2

The concept of convolutional neural networks (CNNs) dates back to the 1980s. Compared to other computer vision methods, they can be trained fully autonomously and require little pre-processing of images which makes them highly applicable for various machine learning tasks. It was only with the introduction of graphics processing units (GPUs) for training CNNs that we saw an explosion of use due to a speed-up in training time through parallelization.

In this chapter, we look at a number of application fields and describe the state of the art, open issues, and future challenges for research involving convolutional neural networks. The rest of the chapter is structured as follows:

- Section 2 introduces convolutional neural networks.
- In Sect. 3 we look at the application of CNNs for object detection, the most common problem solved using convolution.
- Section 4 presents an overview of CNNs for segmentation, or the division of images into regions.
- Section 5 describes how CNNs can be used in reinforcement learning.
- In Sect. 6 we turn our attention to how we can train CNNs using methods of few-shot learning in cases where a large amount of data is not available.
- Section 7 deals with an emerging issue with the use of CNNs, namely the problem of explainability and interpretability.
- Section 8 concludes the chapter.

2 Convolutional Neural Networks

FNNs use general matrix multiplication to propagate signal from input to output. CNNs are neural networks that use convolution instead of general matrix multiplication in at least one of their layers [1, chap 2.].

The idea of sharing weights was first proposed by Fukushima [2], LeCun et al. [3] modified the idea and used convolution for sharing weights in LeNet. CNNs came to prominence with [4], when their GPU implementation and architecture won the object recognition challenge in 2012 on the Imagenet dataset.

A typical CNN for image classification takes grid input with 3 dimensions $height \times width \times n$, where n is usually the number of color channels. The input is then convolved by multiple convolution filters (kernels, masks) with a certain stride. Stride is the step by which the kernel is moved. This produces a feature map whose resolution depends on the parameters of convolution, number of kernels, stride and size of the input image. The feature map is then transformed through an activation function - often rectifier, sigmoid or tanh. Then, the activated input is downsampled by the application of pooling. The output of pooling connects to the next convolutional layer. We apply this process multiple times. Finally, the output of successive convolution layers is passed to fully connected layers for classification.

Using convolution yields important results [1, chap 9.]:

- **Sparse interaction** – accomplished by making the kernel smaller than the input. This sparseness allows us to detect features that occupy fractions of the input.

- **Parameter sharing** – refers to the application of a kernel to each element in convolution (pixel in the image), in contrast with fully connected neural networks, where every weight maps to exactly one output. This sharing further reduces the computational costs of CNNs.
- **Equivariance to translation** – deterministic weight sharing of convolution results in the same translation in both the input and output feature maps. An object moves in a 2D grid the same way than its feature map does.

3 2D/3D Object Classification

CNN-based models are often used in the cognitive science literature as an approach to modeling the behavior of neural systems that perform complex cognitive operations like reasoning and classification.

In computer vision, CNNs have been demonstrated as an interesting technique for making predictions and they have been used to predict the distribution of information about a particular object.

The following subsections present the most commonly used CNN techniques for classification of 2D and 3D images.

3.1 2D Classification

In 1998 Yann LeCun et al. [3] presented the first convolutional neural network called LeNet (see architecture on Fig. 1). The work was dedicated to identifying handwritten digits for zip code recognition in the postal service.

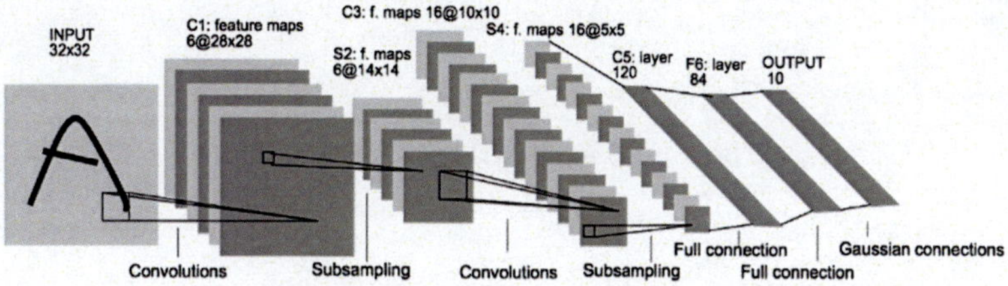

Fig. 1. Architecture of LeNet-5 [3].

After that in 2012, the winners of ImageNet Large-Scale visual Recognition challenge [5] proposed a convolutional model for classification. It achieved a top 5 test error rate of 15.4% with the next best entry achieving an error of 26.2% [4]. In their architecture, the authors used the Rectified Linear Unit Function (ReLU) [6] for the nonlinearity functions which led to a decrease in training time compared to the conventional tanh function. They also used data augmentation techniques that consisted of image translations, horizontal reflections, and patch extractions, and implemented dropout layers in order to solve the problem of overfitting. They trained the model using batch stochastic gradient descent, with specific values for momentum and weight decay.

In 2014, Inception was introduced by Google. The model used deep convolutional neural network architecture [7]. they used inceptions which allowed the use of a convolution layer, size of kernel, type of pooling and concatenate all the outputs and let the network learn better.

In the same year, a model from the University of Oxford was proposed by Simonyan et al. [8]. This model achieved 92.7% top-5 test accuracy on the ImageNet dataset. It was one of the famous models submitted to ILSVRC-2014. It makes the improvement over AlexNet [4] by replacing large kernel-sized filters (11 and 5 in the first and second convolutional layer, respectively) with multiple 3×3 kernel-sized filters one after another (Fig. 2).

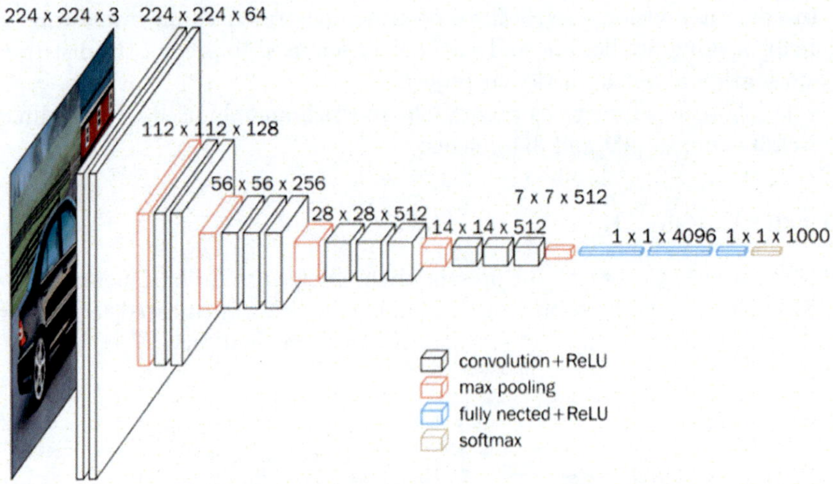

Fig. 2. The architecture of VGG16 [8].

3.2 3D Classification

Most research on classification utilized 2D static images or 2D dynamic videos. Recently, 3D static or dynamic data has received increasing attention due to its explicit representation of geometric structures and their inherent capacity to handle facial pose and illumination variations. Similar to the 2D images, 3D modality can also be represented in static space and dynamic space.

- **Point Cloud based classification**
  Xu et al. [9] proposed SpiderCNN, an extension of convolutional operations from regular grids to irregular point sets which can be embedded after parameterizing convolutional filters. These filters are designed to capture the local

geometric information and Taylor polynomial. SpiderCNN extracts semantic deep features, since it inherits the multi-scale hierarchical architecture from classical CNNs.
Li et al. [10] addressed the problem of desertion of shape information caused by the irregular and unordered point clouds, thus the authors proposed PointCNN, a generalized CNN for leveraging spatially-local correlation from data represented as a point cloud.

- **Multi-view Images based classification**
Su et al. [11] proposed a multi-view CNN for 3D shape recognition from a collection of rendered views on 2D images. The architecture (see Fig. 3) offers better recognition performance since it combines information from multiple views of a 3D shape into a compact descriptor.
Shi et al. [12] proposed DeepPano, which is a rotation invariant deep representation for 3D shape classification. It reconstructs cylinder projection (panoramic view) from 3D shapes, then features are learnt using a deep CNN. Their architecture outperformed other methods by a large margin.
- **Volumetric-based classification**
Maturana et al. [13] proposed VoxNet, a voxel-based classification architecture that uses volumic occupancy grids from LiDAR and RGBD as an input after, they integrate the volumetric representation with a supervised 3D Convolutional Neural Network (3D CNN)
Wang et al. [14] presented a CNN-based architecture using sparsed octree representation of 3D shapes as well as local orientation of the shape that helps to compact storage as well as fast computation.

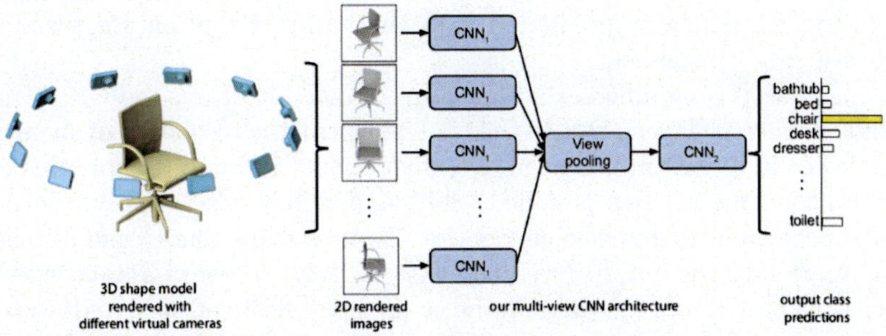

Fig. 3. Multi-view CNN for 3D shape recognition (illustrated using the 1st camera setup). At test time a 3D shape is rendered from 12 different views and is passed through CNN1 to extract view based features. These are then pooled across views and passed through CNN2 to obtain a compact shape descriptor [11].

4 Image Segmentation

Segmentation is a crucial step in image processing tasks. It aims to divide an image into regions with specific properties. Pedestrian localization [15], medical image analysis ([16,17]), scene understanding for autonomous driving [18] are just a few examples where DL-based segmentation is a building block for a higher abstraction system. There are three types of segmentation:

Semantic segmentation – assigns a class label to each pixel from a list of available classes that usually represent different types of objects e.g. background, human. Since we are assigning a value to each pixel, we refer to semantic segmentation as dense prediction.

Instance segmentation – assigns a new class for each occurrence of an object e.g. car, 1st human, 2nd human. It does not assign class labels to each pixel.

Panoptic segmentation [19] – combines semantic and instance segmentation. Every pixel in the image has a class label, while every occurrence of an object is distinguished.

4.1 Semantic Segmentation

We can perform semantic segmentation with CNNs with minor modifications. CNNs are a state-of-the-art solution for complex segmentation tasks. Typical recognition networks such as AlexNet [4], VGGNet [8] or ResNets [20] map input with defined resolution to a vector, which represents an affinity to a certain class. Fully convolutional networks (FCNs) can process an input of any size and produce the corresponding resampled output. The loss function is then composed as a sum over spatial dimensions. Networks designed in this way can be trained end-to-end for pixelwise prediction with the backpropagation algorithm using stochastic gradient descent.

Long et al. [21] modified existing recognition architectures to FCN by removing fully connected layers. Due to the pooling layers, the resolution of an output is much lower than the input image. Deconvolutional layers upsample this coarse output [22]. This process is called 'Shift and stitch'. Specifically, the output layer is connected to the various depth of feature maps (shift) and stitched by deconvolution to the output (stitch). FCNs entirely consist of a layer with three dimensions of $h \times w \times d$, where h and w are spatial dimensions and d is either the color or the feature channel. FCNs build on equivariance to the translation of their components – convolution, pooling, activation function – and work with a local input region. Standard FCNs contain only skip connection from the contractive path to recover spatial information lost during downsampling. [23] extended FCN by addition of short skip connections, that are similar to ones used in residual networks. This approach yielded a significant improvement in gradient backpropagation. The authors reported that short skip connections allow for faster convergence when training deep models.

Ronnenberger et al. [24] developed U-Net for precise segmentation in biomedical imaging (see Fig. 4). The proposed architecture builds on top of FCNs [21].

The main idea behind FCNs is the supplementation of contraction layers with expansive layers to localize high-level spatial features of an image. Similarly, U-Net propagates information through a contractive and expansive path, with skip connection for fine resolution. The main shortcoming of the FCN architecture is the size of the training image corpus required to build a robust and precise model. U-Net incorporates several modifications to deal with shortcomings of FCNs. After down-scaling, the network propagates information at a sufficient bandwidth to pass complex information to higher resolution layers. This composition does not require any fully connected layers and only uses non-zero padded convolution. U-net and its 3D versions [24–26] calculate the error for each pixel as weighted cross-entropy.

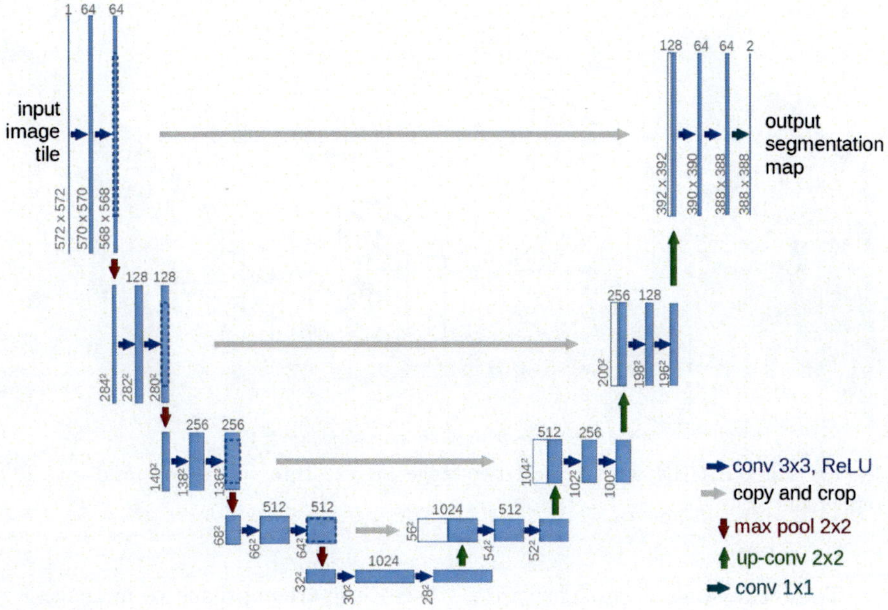

Fig. 4. The U-Net architecture that contains two paths contractive – left, expansive – right, forming a typical U-shape [24]

4.2 Instance Segmentation

Following the success of CNNs in image classification and object detection, another approach performing segmentation was developed - instance segmentation. It builds on top of object detection networks such as Fast R-CNN [27], that use a sliding window to detect objects in an image. Feature extraction is commonly performed by ResNets or VGG, referred to as a backbone of the architecture. The output of this CNN is shared between building blocks across the segmentation architecture. Specifically, Faster R-CNN [28] contains a Region Proposal Network (RPN) which directly uses a shared output of CNN for proposing the region of interest (ROI) that contains the recognizable object. The output of the RPN is then forwarded to the ROI pooling layer, which by performing successive pooling selects the regions for the fully connected layers. These fully connected layers then output the class label and the bounding box. Mask R-CNN [29] (Fig. 5) adds a binary mask for each rectangle of interest that refines the bounding box of an object, therefore performs instance segmentation. YOLACT [30] achieves real-time instance segmentation through parallel computation of a set of prototype masks and predicting per instance mask coefficients. The final mask is generated as a linear combination of the prototype mask and mask coefficients.

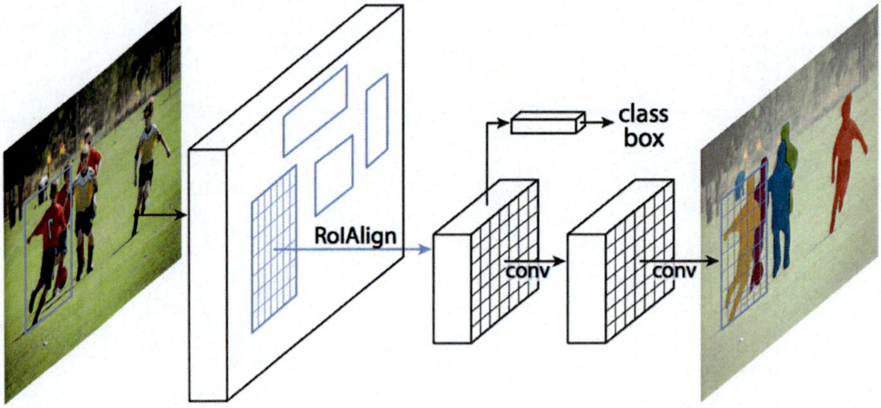

Fig. 5. Mask R-CNN with typical two-stage architecture - ROI proposal and ROI refinement [29]

These two-stage approaches are a state-of-the-art approach to instance segmentation. The foundations of dense mask predictions (one-stage instance segmentation) are presented in [31]. They formalize the task as 4D Tensor prediction and with their CNN TensorMask demonstrate results comparable to Mask R-CNN.

4.3 Advanced Modifications for Feature Extraction

Jegou et al. [32] extended the idea behind DenseNets [33] to perform semantic segmentation. They achieved state-of-the-art results on benchmark datasets such as CamVid and Gatech, without pretraining and post-processing. Furthermore, due to its architecture, their model contains fewer parameters than the following models on benchmark tests.

Chen et al. [34] developed CUMedVision, a deep contextual network that leverages multilevel contextual information from the deep hierarchical structure to achieve better segmentation performance. To further improve the robustness against vanishing gradients and strengthen the capability of the backpropagation of gradient flow, they incorporate auxiliary classifiers in the architecture of their deep neural network. They outperformed the state-of-the-art solution on the benchmark dataset 2012 ISBI segmentation challenge of neuronal structures. The architectures DeepLab v1–v3 [35,36] use dilated/atrous convolutions to enlarge the field of view of filters to incorporate a larger context. Atrous convolution is an efficient mechanism to control the field of view within a layer with accurate localization of pixels and spatial context. For post-processing, they use fully-connected Conditional Random Field (CRF). CRF overcomes limitations of previous models as it can handle arbitrarily large neighborhoods while preserving fast inference times [37].

5 Reinforcement Learning

Reinforcement learning (RL) [38] is a machine learning approach in which knowledge is not gained from a dataset, but from experience. Formally, RL can be described as an interaction between an agent and an environment. The agent has a goal that it wants to accomplish through selecting actions that might cause the environment to change its state. RL methods then use this change as an indicator for the agent to determine how suitable the selected action was given the agent's goal. The indicator is also expressed as a numerical reward, defined by the reward function, a mathematical function mapping state changes to numerical rewards. The higher the reward, the more appropriate the action selected. Throughout training, the agent tries to maximize its cumulative reward over the course of the interaction.

In reinforcement learning, choosing the proper reward function and state representation is crucial to the success of training. While there are a number of guidelines for defining useful reward functions, state representation remained a large problem, since basic reinforcement learning algorithms work well only with discrete state spaces. An application of deep learning in reinforcement learning led to RL algorithms that are able to deal with continuous state spaces [39], while introducing convolutional neural networks enables the derivation of suitable state representations.

5.1 Reinforcement Learning in Computer Games

One of the first and best-known instances of convolutional neural networks being used in a combination with RL was Deep Q-network (DQN) [39]. It was successfully trained to play Atari 2600 games (including Breakout) directly from frames that would be displayed on screen.

The input image is converted from RGB to grayscale to reduce dimensionality, and cropped so the resulting image covers only the rough playing area. Since it is impossible to infer direction and speed from a single static image, four consecutive images are stacked to form the input to the network. The input

($84 \times 84 \times 4$) is processed by two convolutional layers (16 8×8 filters with a stride of 4 and 32 4×4 with a stride of 2 respectively), each followed by a rectifier and a final hidden fully-connected layer with 256 neurons. The output ranges between 4 and 18 neurons based on the game played and represents the expected value of executing one of the available actions in the given environment state (Fig. 6). It is important to note that the convolutional layers play no role in the action selection process, they only transform the state representation.

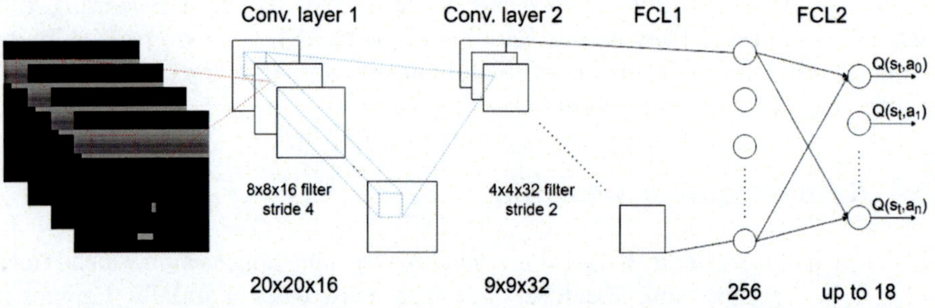

Fig. 6. Deep Q network topology used in playing Atari games. Each of the two convolutional layers is followed by a rectifier. The network outputs the predicted Q-value for each valid action. Based on [39].

Optimization algorithms assume the data is uniformly and independently distributed, which might not hold true in the case of RL. To address this problem, samples obtained from a sequence of actions are stored in a buffer. The buffer – called replay memory – stores the agent's experiences pooled over multiple episodes and transformed into fixed-length representation. This allows for the use of randomly sampled past experiences for mini-batch learning. Afterwards, the network can be trained in a supervised manner.

By adding an LSTM layer [40] and creating a Deep Recurrent Q-network (DRQN) [41], we can avoid frame stacking. This approach has been successfully applied in 3D first-person shooters, such as Doom [42] (based on the VizDoom platform [43]).

Rather than using the actual image output (that would be shown to the player), some board-game algorithms, such as AlphaGo [44], form their own visual representation of the board.

5.2 Reinforcement Learning for Object Detection

A combination of reinforcement learning and CNNs can also be used for object localization as shown in [45]. While standard CNNs first look at smaller sections of an image to detect objects, this approach does not represent the way humans do it. Tree-structured reinforcement learning (Tree-RL) first considers the entire image on which objects should be detected, and then makes a sequence of decisions to detect multiple objects, taking into account their interdependence. The

RL agent in this method has two actions at its disposal: it can either translate the current search window, or it can scale it to a different size. This way the agent learns multiple near-optimal policies that enable it to better cover objects with a great variety. The effectiveness of the method was validated on the PASCAL VOC 2007 and 2012 datasets where it achieved comparable results to state-of-the-art detection systems.

Another research topic in computer vision where reinforcement learning and convolutional neural networks show promising potential is point cloud parsing, i.e. recognizing objects from 3D data. Lui et al. [46] present a method that utilizes a 3D CNN, a deep Q-network (DQN) and a residual recurrent neural network (RRNN) to achieve good performance on point cloud semantic parsing. Similarly to Tree-RL, this method also models the way humans recognize objects, in that it provides an eye window to enable the CNN to recognize an object from the point cloud. While the CNN is responsible for learning basic object features (such as shape, color, and context of the points), the DQN adjusts the size and position of the eye window. The two networks cooperate to determine the size and position of the eye window most suitable to object detection, and finally the RRNN provides the classification result.

5.3 Reinforcement Learning in Robotics

RL algorithms usually require a large number of interactions to achieve high performance, which in turn is only possible with environments that are easy to model, but neither is available in real-life scenarios. This problem arises primarily in robotics applications, where reinforcement learning has been proven to be a promising approach. Robots are often used in environments that are hard to model, or they have to work with or close to humans, which makes it critical for the learning agent not to make mistakes during learning.

The problem was addressed in [47] where simulations were used to train a primary policy and then the robot adapted to a real-world environment using reinforcement learning. The goal of the robot was to reach a target position, the steps to which were learned in simulation. A CNN was used for object detection, that is to track the target object, providing input to the reinforcement learning agent. The approach was tested in experimental setups, where the robot accurately located and reached the target position. The method is highly flexible and can be adapted to various simple tasks that are to be carried out in dynamic environments.

Manipulation is a task often addressed in robotics, and the newest research utilizes reinforcement learning to enable the robot to learn the proper behavior from scratch. However, the robot must observe its environment in a way that provides the necessary information for it to achieve its goal. CNNs can provide this information, as it was shown in [48]. The authors of the paper trained a robot that combined two standard actions, pushing and grasping, which are usually studied in isolation. The robot used deep reinforcement learning to learn both pushing and grasping policies and to use them in combination to achieve its goals. Input to the network were visual observations, which were processed using convolution, and the output of the network were the expected Q-values representing the utility of the individual actions.

6 Few-Shot Learning

Although recent CNN based methods have achieved tremendous success in multiple domains within supervised learning settings, the performance of these methods deteriorates drastically in the absence of densely available labels, unlike for humans who can generalize incrementally to novel classes by observing only a few examples [49]. Although many large and diverse labeled datasets exist, there are situations in which only a limited amount of "samples per concept" is available, causing the dilemma of recognizing common patterns particularly laborious. When only a limited amount of data is available, training a function able to generalize becomes much harder. In fact, rather than finding meaningful patterns related to the task, a function's expressive capacity can end up being used to model certain traits of the data that have limited or no causal relationship with the desired output. This phenomenon is commonly called overfitting. Few-shot learning tries to solve these problems.

Few-Shot Learning (FSL) is, therefore, a type of machine learning problem, where the available experience contains only a limited number of examples with *(weakly)* supervised information for the target task. A particular example of few-shot learning is few-shot classification. It is a N-shot K-way classification problem, where N is the number of labeled samples per class, and K is the number of classes to classify among. Therefore our training data contains $I = NK$ examples [50]. However, this paradigm applies to any problems *i.e.* few-shot regression [51], few-shot object-detection [52–55], few-shot segmentation [56–59], and few-shot reinforcement learning (Meta Reinforcement Learning) [60–63].

Zero-shot learning (ZSL) is a variant of few-shot learning problem where no training data is available for some of the classes. Most ZSL algorithms use some connection between the available information and the unseen classes. In ZSL, there are some labeled training instances in the feature space. The classes covered by these training instances are referred to as seen classes [64]. In the feature space, there are also some unlabeled testing instances, which belong to another set of classes. These classes are referred to as unseen classes. ZSL techniques are still in their infancy and are a relatively active topic in research. The ZSL setting can be tackled by solving related sub-problems, *e.g.* learning intermediate attribute classifiers [65] and learning a mixture of seen class proportions [66–68], or by a direct approach, *e.g.* compatibility learning frameworks [69–72].

Multiple methods can be used to solve this FSL/ZSL setting. Such methods are data augmentation or hallucinations [73–77], and meta learning [78,79].

6.1 Hallucination-Based Methods

Neural networks suffer from over-fitting and catastrophic forgetting when trained with a limited amount of data. A natural way to solve this problem is to use data augmentation. However, standard data augmentation such as rotation, image translations, horizontal reflections, produces only limited plausible alternative data. Hallucination-based methods use a hallucinator to generate new samples. The hallucinator is generally a generative model. It takes a single example and produces other examples in different poses, different background or different surrounding [74]. Antoniou et al. [73] and Gao et al. [75] designed a generative model to do data augmentation. It takes data from a source domain and learns to take any data item and generate other within-class data items.

6.2 Meta-Learning

Meta-learning was introduced as a framework to address the few-shot learning setting [80]. Unlike the conventional machine learning approach, meta-learning is performed on a set of tasks. Each task has its own training and test set. The key concept is to use a variety of similar few-shot tasks to learn how to adapt a base-learner to a new task for which only a few labeled samples are available. Those tasks may include potential unseen tasks. In other words, the meta-learning framework *learns how to learn* given a set of training tasks and evaluates itself using a set of test tasks. This is mainly motivated by the human ability to leverage prior experiences when tackling a new task. There are three common approaches to meta-learning: metric-based, model-based, and optimization-based. In this review, we only focus on metric-based meta-learning algorithms.

Metric Learning or metric based meta-learning algorithms are mainly similar to nearest neighbor algorithms and/or kernel density estimation. Fundamentally, they learn a distance function over objects. Therefore, they automatically construct task-specific distance metrics from *(weakly)* supervised data. The learned distance metric is afterward used to perform various tasks *e.g.* k-nearest neighbors classification, clustering, information retrieval.

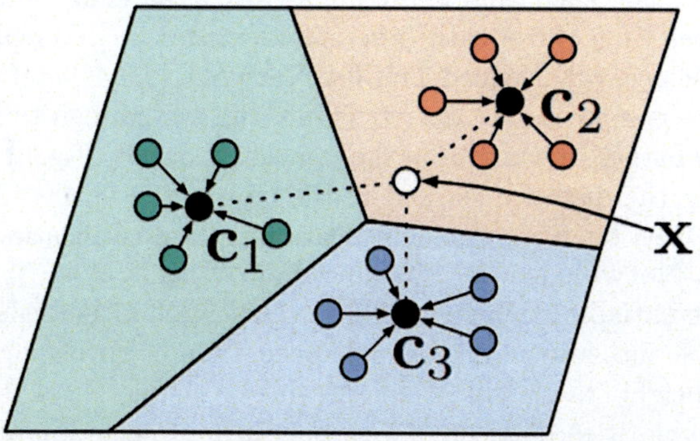

Fig. 7. Prototypical Network in few-shot learning [81].

- **Prototypical Network** (ProtoNet) was first introduced by Snell et al. [81]. It learns class prototypes from a high-level description of data. Those data can be labeled data or natural language description. In other words, this network learns a metric space in which classification can be performed by computing distances to prototype representations of each class [81]. The intuition behind this method is that there exists an embedding space in which samples from each class cluster around a single prototypical representation which is simply the mean of the individual samples (Fig. 7). Although multiple distance metrics can be used with Prototypical Network, squared Euclidean distance is currently known to work best.
  However, the Prototypical Network performs poorly when the support set has high variance. Mukaiyama et al. [82] combined ProtoNet with local Fisher discriminant analysis to reduce the local within-class covariance and increase the local between-class covariance of the support set.
- **Siamese Neural Network** (SNN) was first introduced in 1995 by Bromley et al. [83]. Later Koch et al. [84] decided to use it for one-shot classification. Primarily, as shown in Fig. 8, it takes two separate examples as inputs instead of just one. Each of these is mapped from a high-dimensional input space to a low-dimensional space by an encoder network. Most of the time, a CNN is used as an encoder of the network. The two encoder networks are 'twins' because they share the same weights and learn the same function. These two networks are then joined by a layer that measures the distance (*e.g.* the Euclidean distance, the cosine) between the two samples in the embedding space. The network is therefore trained to minimize this distance for similar samples and to amplify it for disparate samples. In other words, it learns feature embedding so that the distances capture **the semantic similarity**. By including this distance layer, it is possible to optimize the properties of the embedding space directly instead of optimizing for classification accuracy.
  SNN can also be trained using the triplet loss function [85,86]. This loss function operates on 3 inputs: the anchor which is an arbitrary data point, a positive sample that belongs to the same class as the anchor, and a negative sample that belongs to a different class from the anchor. It optimizes the embedding space such that data points with the same identity are closer to each other than those with different identities [87]. When using this loss function, the network is called **Triplet Network** [88]. Therefore it is based on two pairwise comparisons. After training, the system can take two examples and verify whether they are from the same class or not (Fig. 9). This is done by thresholding the distance learned in the embedding space. Triplet Networks are mainly used for person reidentification or face verification.
- **Matching Networks** were introduced by Vinyals et al. [89]. It is an end to end differentiable nearest neighbor. This method embeds a high dimensional sample into a low dimensional space, then performs a generalized form of nearest-neighbors classification. As shown in Fig. 10, Matching Networks compute similarities between the embeddings of each support example and the query example. However, this has some disadvantages: the algorithm is not robust to data imbalance; if there are more support examples for some

classes than others (*i.e.*, we have departed from the N-way-K-shot scenario), the ones with more frequent training data may dominate.

- **Relation Networks** proposed by Sung et al. [90] also learn a metric for comparison of the embeddings. Similarly to prototypical networks, the relation network averages the embeddings of each class in the support set to form a single prototype. Each prototype is then concatenated with the query embedding and passed to a relation module. This is a trainable non-linear operator, generally CNN, that produces a similarity score between 0 and 1 where 1 indicates that the query example belongs to this class prototype. This approach can be trained end-to-end.

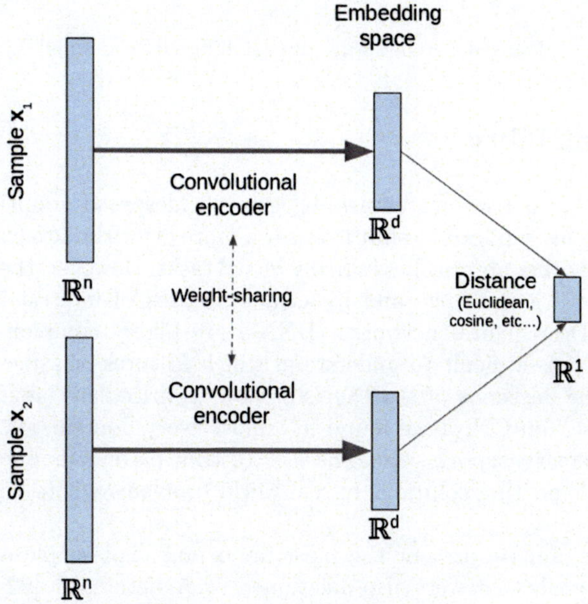

Fig. 8. An example of Siamese Neural Network [84].

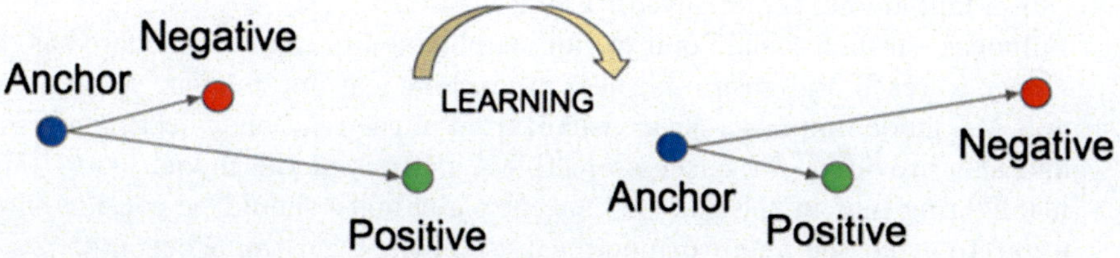

Fig. 9. Learning procedure of a Triplet Network [88].

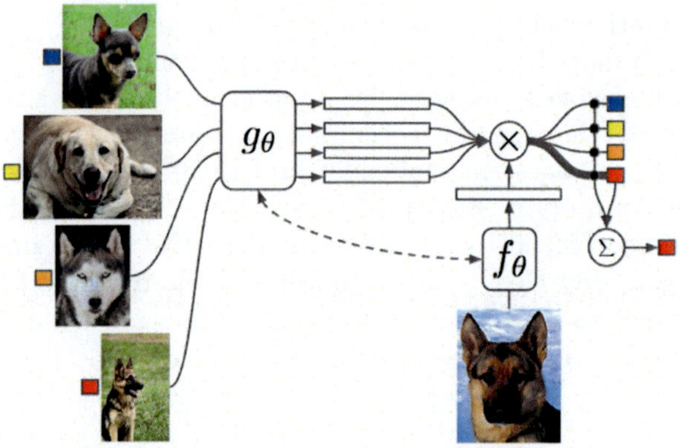

Fig. 10. Architecture of Matching Networks [89].

7 Explainability

In our everyday life we are witnessing a fast widespread adoption of artificial intelligence. This can lead to shift towards a more algorithmic society [91]. CNNs outperformed other approaches in many visual tasks. However, the interpretation of CNNs is still a challenge and it became an area of interest in the research community. Deep neural networks (DNNs) are highly complex and nonlinear functions, so it is difficult to understand which features or aspects of the input data affect the decisions of the DNNs or how a particular classification occurs. Solutions based on CNNs are found in almost every domain, such as the army, healthcare, security, sports, etc. The lack of transparency is one of the critical issues in implementing solutions in real-world problems where a mistake can be costly.

In this section we present the most important explainable artificial intelligence (XAI) goals that were also mentioned by Arrieta et al. [92] and the traits of an explanation defined by Xie et al. [93]:

- **Causality** – finding causality among data variables is a common goal for explainability. Some authors argue that explainable models might simplify the task of finding connections that could be tested further for a stronger causal link among the involved variables [94,95].
- **Fairness** – from a social point of view, explainability can be considered as the ability to reach and ensure fairness in machine learning models. An explainable ML model implies a clear visualization of the relations affecting a result and also providing for fairness or ethical analysis of the model [96,97] and highlighting bias in the data [98,99]. Explainability should be considered as a tool to avoid the unfair or unethical use of the algorithm's outputs.
- **Accessibility** – several authors argue for explainability as the property that allows end-users to get more involved in the process of improving and developing a certain ML model [100,101].

- **Trustworthiness** – trustworthy neural networks are neural nets whose decision-making process has no need for validation. Several authors define trustworthiness as a main goal of an XAI model [102,103]. A prediction with high probability does not assure its trustworthiness, as shown in recent adversarial studies [104–106].
- **Confidence** – for a neural network from which reliability is expected, confidence should be always assessed. Stability is a must-have feature when drawing interpretations from a certain model as stated in [107,108]. Models that are not stable can not produce trustworthy interpretations.
- **Safety** – DNNs that may have an impact on human life, wealth, or societal policy should be *safe*. A safe DNN should: (i) consistently perform as expected; (ii) given impulse from its input, guard against decisions that can negatively affect the user or society; (iii) exhibit high reliability under both standard and extraordinary operating circumstances; (iv) provide feedback to a user about how operating conditions affect its decisions [93].
- **Ethics** – A DNN behaves ethically if its decisions and decision-making process does not break a code of ethical policies defined by the user. The correct way to decide if a DNN is acting ethically is a topic for discussion [93].

Human cognitive skills prefer the understanding of visual data over other kinds of models, that is why reaching explainability for CNNs is simpler. Current research papers on understanding CNNs can be divided into two categories: 1) Those that try to explain the decision process by mapping back the output in the input space to see which parts of the input were decisive for the output; and 2) those that try to examine inside the network and understand how the intermediate layers see the external world, not significantly correlated to any specific input, but in common [92]. Adding insights into the decision making process of classifiers is one of the key elements in interpretable AI.

Based on the review [93] there are several types of explainable methods: (i) visualization methods; (ii) distillation methods; (iii) intrinsic methods. In this review we focus only on visualization approaches. Visualization methods display an explanation by highlighting, via a scientific visualization, features of an input that strongly influence the output of a CNN.

7.1 Backpropagation-Based Methods

Backpropagation-based methods are able to visualize feature relevance based on the volume of gradient passed through network layers during network training.

Discovering the right image representations is a critical problem in computer vision. Deconvolution was for the first time defined by Zeiler et al. [109] as an algorithm to learn image features in an unsupervised manner. They presented a hierarchical model that learns image decompositions via alternating layers of convolutional sparse coding and max pooling. Experiments were done with natural images from the Caltech-101 and Caltech-256 datasets. The model captures image features, such as low-level edges, mid-level edge junctions, high-level object parts and complete objects. Zeiler et al. [22] introduced a way to visualize individual features at each layer in the neural network model. This method is also considered to be a perturbation method. By using this visualization technique, it is possible to see the evolution of the features during the training process and to detect potential issues. To visualize these strongest activations in the input image, the same authors used the occlusion sensitivity approach to create a saliency map, which consists of iteratively forwarding the same image through the network occluding a different area at a time.

Transparency and trustworthiness of CNN models is supported by visualization of the Gradient-weighted Class Activation Mapping [110] technique which uses the gradients of any target concept, moving into the final convolutional layer to produce a coarse localization map, highlighting the important areas in the image for predicting the concept. Grand-CAM also improves interpretability and faithfulness to the original model.

To improve the quality of the mapping on the input space in [111] the authors recommended simplifying both the CNN architecture and the visualization method. Zhou et al. [111] included a global average pooling layer between the last convolutional layer of the CNN and the fully-connected layer that predicts the object class. With this simple structural change of the CNN, the authors built a class activation map that helps classify the image regions that were especially important for a specific object class by projecting back the weights of the output layer on the convolutional feature maps.

The LRP (Layer-Wise Relevance Propagation) technique was first used for natural language processing in [112]. Arras et al. [113] tried to describe the predictions of CNNS on a topic categorization task by highlighting the words which were decisive for a specific prediction. Word deleting methods were used in the experimental part. The results showed that LRP outperformed sensitivity analysis.

An approach for computing importance scores in a multi-layer neural network called DeepLIFT (Deep Learning Important FeaTures) was introduced by Shrikumar et al. [114]. Their technique compares the activation of a node to the reference activation and assigns the score according to the difference. DeepLIFT decomposes the output prediction of a neural network on a specific input by backpropagating the contributions of all neurons in the network to every feature of the input.

7.2 Perturbation-Based Methods

Perturbation-based methods compute input feature importance by changing or removing the input feature. After this they compare the difference in network output between the original and adjusted ones. Perturbation methods can estimate the marginal relevance of each feature with regards to how a network responds to a particular input.

Via visualization of the decision process in neural network we can reach higher interpretability and afterwards trustworthiness. It can accelerate the adoption of black-box classifiers. Zintgraf et al. [115] proposed a technique whose output is a visualization of how the neural network responds to a specific input in order to explain a particular classification made by the network. By removing information from the image and evaluating the effect of this it was possible to highlight parts of the input image that support the decision of a neural network for or against a certain class. Experiments were done on natural images (ImageNet data), as well as medical images (MRI brain scans).

8 Conclusion

In this chapter, we described CNNs and their application in various fields of research. In Sects. 3 and 4 we reviewed how CNNs can be used for object detection in 2D and 3D data, and for image segmentation. We presented in Sect. 5 how the addition of convolution to deep reinforcement learning methods can solve the problem of state representation.

While CNNs have a wider range of application than simple feed-forward neural networks, especially with regards to processing spatial data, their application is also limited. We looked at two problems usually associated with CNNs. First, we described in Sect. 6 how a lack of training data can make it hard to apply convolutional neural networks, and how few-shot learning can address this problem. In Sect. 7, we looked at the explainability and interpretability of convolutional neural networks, which is a highly desirable feature as deep learning is applied in more complex domains.

Convolutional neural networks are a well-researched topic in machine learning due to their high performance and versatility. They revolutionized computer vision, and have found their application potential in almost all domains. Due to their ability to discern spatial information, they widened the range of data that is processible by machine learning algorithms. However, as they are applied in more and more complex tasks, their inherent limitations are becoming apparent, and there is a need to combine them with other methods of artificial intelligence to overcome these weaknesses.

References

1. Goodfellow, I., Bengio, Y., Courville, A.: Deep Learning. MIT Press (2016). http://www.deeplearningbook.org
2. Fukushima, K.: Neocognitron: a self-organizing neural network model for a mechanism of pattern recognition unaffected by shift in position. Biol. Cybern. **36**(4), 193–202 (1980)
3. LeCun, Y., Bottou, L., Bengio, Y., Haffner, P., et al.: Gradient-based learning applied to document recognition. Proc. IEEE **86**(11), 2278–2324 (1998)
4. Krizhevsky, A., Sutskever, I., Hinton, G.E.: Imagenet classification with deep convolutional neural networks. In: Advances in Neural Information Processing Systems, pp. 1097–1105 (2012)
5. Stanford Vision Lab Stanford University, P.U.: (imagenet large-scale visual recognition challenge). http://www.image-net.org/
6. Nair, V., Hinton, G.E.: Rectified linear units improve restricted Boltzmann machines. In: Proceedings of the 27th International Conference on Machine Learning (ICML 2010), pp. 807–814 (2010)

7. Szegedy, C.,Liu, W., Jia, Y., Sermanet, P., Reed, S., Anguelov, D., Erhan, D., Vanhoucke, V., Rabinovich, A., et al.: Going deeper with convolutions. arxiv 2014. arXiv preprint arXiv:1409.4842, **1409** (2014)
8. Simonyan, K., Zisserman, A.: Very deep convolutional networks for large-scale image recognition. arXiv preprint arXiv:1409.1556 (2014)
9. Xu, Y., Fan, T., Xu, M., Zeng, L., Qiao, Y.: SpiderCNN: deep learning on point sets with parameterized convolutional filters. arXiv preprint arXiv:1803.11527 (2018)
10. Li, Y., Bu, R., Sun, M., Wu, W., Di, X., Chen, B.: PointCNN: convolution on x-transformed points. In: Advances in Neural Information Processing Systems, pp. 820–830 (2018)
11. Su, H., Maji, S., Kalogerakis, E., Learned-Miller, E.: Multi-view convolutional neural networks for 3d shape recognition. In: Proceedings of the IEEE International Conference on Computer Vision, pp. 945–953 (2015)
12. Shi, B., Bai, S., Zhou, Z., Bai, X.: DeepPano: deep panoramic representation for 3-d shape recognition. IEEE Signal Process. Lett. **22**(12), 2339–2343 (2015)
13. Maturana, D., Scherer, S.: Voxnet: A 3D convolutional neural network for real-time object recognition. In: 2015 IEEE/RSJ International Conference on Intelligent Robots and Systems (IROS), pp. 922–928. IEEE (2015)
14. Wang, P.S., Liu, Y., Guo, Y.X., Sun, C.Y., Tong, X.: O-CNN: octree-based convolutional neural networks for 3D shape analysis. ACM Trans. Graphics (TOG) **36**(4), 1–11 (2017)
15. Yu, Y., Makihara, Y., Yagi, Y.: Pedestrian segmentation based on a Spatio-temporally consistent graph-cut with optimal transport. IPSJ Trans. Comput. Vision Appl. **11**(1), 10 (2019)
16. Litjens, G., Kooi, T., Bejnordi, B.E., Setio, A.A.A., Ciompi, F., Ghafoorian, M., Van Der Laak, J.A., Van Ginneken, B., Sanchez, C.I.: A survey on deep learning in medical image analysis. Med. Image Anal. **42**, 60–88 (2017)
17. Shen, D., Wu, G., Suk, H.I.: Deep learning in medical image analysis. Annu. Rev. Biomed. Eng. **19**, 221–248 (2017)
18. Chen, C., Seff, A., Kornhauser, A., Xiao, J.: Deepdriving: learning affordance for direct perception in autonomous driving. In: Proceedings of the IEEE International Conference on Computer Vision, pp. 2722–2730 (2015)
19. Kirillov, A., He, K., Girshick, R., Rother, C., Dollár, P.: Panoptic segmentation. In: Proceedings of the IEEE Conference on Computer Vision and Pattern Recognition, pp. 9404–9413 (2019)
20. He, K., Zhang, X., Ren, S., Sun, J.: Deep residual learning for image recognition. In: Proceedings of the IEEE Conference on Computer Vision and Pattern Recognition, pp. 770–778 (2016)
21. Long, J., Shelhamer, E., Darrell, T.: Fully convolutional networks for semantic segmentation. In: Proceedings of the IEEE Conference on Computer Vision and Pattern Recognition, pp. 3431–3440 (2015)
22. Zeiler, M.D., Fergus, R.: Visualizing and understanding convolutional networks. In: European Conference on Computer Vision, pp. 818–833. Springer (2014)
23. Drozdzal, M., Vorontsov, E., Chartrand, G., Kadoury, S., Pal, C.: The importance of skip connections in biomedical image segmentation. In: Deep Learning and Data Labeling for Medical Applications, pp. 179–187. Springer (2016)
24. Ronneberger, O., Fischer, P., Brox, T.: U-net: Convolutional networks for biomedical image segmentation. In: International Conference on Medical Image Computing and Computer-Assisted Intervention, pp. 234–241. Springer (2015)

25. Çiçek, Ö., Abdulkadir, A., Lienkamp, S.S., Brox, T., Ronneberger, O.: 3D u-net: learning dense volumetric segmentation from sparse annotation. In: International Conference on Medical Image Computing and Computer-Assisted Intervention, pp. 424–432. Springer (2016)
26. Milletari, F., Navab, N., Ahmadi, S.A.: V-net: Fully convolutional neural networks for volumetric medical image segmentation. In: 2016 Fourth International Conference on 3D Vision (3DV), pp. 565–571. IEEE (2016)
27. Girshick, R.: Fast R-CNN. In: Proceedings of the IEEE International Conference on Computer Vision, pp. 1440–1448 (2015)
28. Ren, S., He, K., Girshick, R., Sun, J.: Faster R-CNN: towards real-time object detection with region proposal networks. In: Advances in Neural Information Processing Systems, pp. 91–99 (2015)
29. He, K., Gkioxari, G., Dollár, P., Girshick, R.: Mask R-CNN. In: Proceedings of the IEEE International Conference on Computer Vision, pp. 2961–2969 (2017)
30. Bolya, D., Zhou, C., Xiao, F., Lee, Y.J.: Yolact: Real-time instance segmentation. In: Proceedings of the IEEE International Conference on Computer Vision, pp. 9157–9166 (2019)
31. Chen, X., Girshick, R., He, K., Dollár, P.: TensorMask: a foundation for dense object segmentation. In: Proceedings of the IEEE International Conference on Computer Vision, pp. 2061–2069 (2019)
32. Jégou, S., Drozdzal, M., Vazquez, D., Romero, A., Bengio, Y.: The one hundred layers tiramisu: Fully convolutional densenets for semantic segmentation. In: Proceedings of the IEEE Conference on Computer Vision and Pattern Recognition Workshops, pp. 11–19 (2017)
33. Huang, G., Liu, Z., Van Der Maaten, L., Weinberger, K.Q.: Densely connected convolutional networks. In: Proceedings of the IEEE Conference on Computer Vision and Pattern Recognition, pp. 4700–4708 (2017)
34. Chen, H., Qi, X., Yu, L., Heng, P.A.: Dcan: deep contour-aware networks for accurate gland segmentation. In: Proceedings of the IEEE Conference on Computer Vision and Pattern Recognition, pp. 2487–2496 (2016)
35. Chen, L.C., Papandreou, G., Kokkinos, I., Murphy, K., Yuille, A.L.: Semantic image segmentation with deep convolutional nets and fully connected CRFs. arXiv preprint arXiv:1412.7062 (2014)
36. Chen, L.C., Papandreou, G., Kokkinos, I., Murphy, K., Yuille, A.L.: Deeplab: Semantic image segmentation with deep convolutional nets, Atrous convolution, and fully connected CRFs. IEEE Trans. Pattern Anal. Mach. Intell. **40**(4), 834–848 (2018)
37. Kamnitsas, K., Ledig, C., Newcombe, V.F., Simpson, J.P., Kane, A.D., Menon, D.K., Rueckert, D., Glocker, B.: Efficient multi-scale 3D CNN with fully connected CRF for accurate brain lesion segmentation. Med. Image Anal. **36**, 61–78 (2017)
38. Sutton, R.S., Barto, A.G.: Reinforcement Learning: An Introduction. MIT Press, Cambridge (2018)
39. Mnih, V., Kavukcuoglu, K., Silver, D., Graves, A., Antonoglou, I., Wierstra, D., Riedmiller, M.: Playing atari with deep reinforcement learning. arXiv preprint arXiv:1312.5602 (2013)
40. Hochreiter, S., Schmidhuber, J.: LSTM can solve hard long time lag problems. In: Advances in Neural Information Processing Systems, pp. 473–479 (1997)
41. Hausknecht, M., Stone, P.: Deep recurrent q-learning for partially observable MDPs. In: 2015 AAAI Fall Symposium Series (2015)

42. Schulze, C., Schulze, M.: ViZDoom: DRQN with prioritized experience replay, double-q learning and snapshot ensembling. In: Proceedings of SAI Intelligent Systems Conference, pp. 1–17. Springer (2018)
43. Kempka, M., Wydmuch, M., Runc, G., Toczek, J., Jaśkowski, W.: ViZDoom: a doom-based AI research platform for visual reinforcement learning. In: 2016 IEEE Conference on Computational Intelligence and Games (CIG), pp. 1–8. IEEE (2016)
44. Silver, D., Huang, A., Maddison, C.J., Guez, A., Sifre, L., Van Den Driessche, G., Schrittwieser, J., Antonoglou, I., Panneershelvam, V., Lanctot, M., et al.: Mastering the game of go with deep neural networks and tree search. Nature **529**(7587), 484–489 (2016)
45. Jie, Z., Liang, X., Feng, J., Jin, X., Lu, W., Yan, S.: Tree-structured reinforcement learning for sequential object localization. In: Advances in Neural Information Processing Systems, pp. 127–135 (2016)
46. Liu, F., Li, S., Zhang, L., Zhou, C., Ye, R., Wang, Y., Lu, J.: 3DCNN-DQN-RNN: a deep reinforcement learning framework for semantic parsing of large-scale 3d point clouds. In: Proceedings of the IEEE International Conference on Computer Vision, pp. 5678–5687 (2017)
47. Chen, C., Li, H.Y., Dharmawan, A.G., Ismail, K., Liu, X., Tan, U.X.: Robot control in human environment using deep reinforcement learning and convolutional neural network. In: 2019 IEEE International Conference on Robotics and Biomimetics (ROBIO), pp. 1121–1126. IEEE (2019)
48. Zeng, A., Song, S., Welker, S., Lee, J., Rodriguez, A., Funkhouser, T.: Learning synergies between pushing and grasping with self-supervised deep reinforcement learning. In: 2018 IEEE/RSJ International Conference on Intelligent Robots and Systems (IROS), pp. 4238–4245. IEEE (2018)
49. Lake, B.M., Ullman, T.D., Tenenbaum, J.B., Gershman, S.J.: Building machines that learn and think like people. Behav. Brain Sci. **40** (2017)
50. Gidaris, S., Komodakis, N.: Dynamic few-shot visual learning without forgetting. In: Proceedings of the IEEE Conference on Computer Vision and Pattern Recognition, pp. 4367–4375 (2018)
51. Loo, Y., Lim, S.K., Roig, G., Cheung, N.M.: Few-shot regression via learned basis functions (2019)
52. Wang, X., Huang, T.E., Darrell, T., Gonzalez, J.E., Yu, F.: Frustratingly simple few-shot object detection. arXiv preprint arXiv:2003.06957 (2020)
53. Fan, Q., Zhuo, W., Tang, C.K., Tai, Y.W.: Few-shot object detection with attention-RPN and multi-relation detector. In: Proceedings of the IEEE/CVF Conference on Computer Vision and Pattern Recognition, pp. 4013–4022 (2020)
54. Yang, Z., Wang, Y., Chen, X., Liu, J., Qiao, Y.: Context-transformer: tackling object confusion for few-shot detection. In: AAAI, pp. 12,653–12,660 (2020)
55. Hsieh, T.I., Lo, Y.C., Chen, H.T., Liu, T.L.: One-shot object detection with co-attention and co-excitation. In: Advances in Neural Information Processing Systems, pp. 2725–2734 (2019)
56. Xie, E., Sun, P., Song, X., Wang, W., Liu, X., Liang, D., Shen, C., Luo, P.: Polarmask: single shot instance segmentation with polar representation. In: Proceedings of the IEEE/CVF Conference on Computer Vision and Pattern Recognition, pp. 12,193–12,202 (2020)
57. Zhang, X., Wei, Y., Yang, Y., Huang, T.S.: SG-ONE: similarity guidance network for one-shot semantic segmentation. IEEE Trans. Cybern. (2020)
58. Park, Y.H., Seo, J., Moon, J.: Cafenet: class-agnostic few-shot edge detection network. arXiv preprint arXiv:2003.08235 (2020)

59. Zhao, Y., Price, B., Cohen, S., Gurari, D.: Objectness-aware one-shot semantic segmentation. arXiv preprint arXiv:2004.02945 (2020)
60. Zintgraf, L., Shiarlis, K., Igl, M., Schulze, S., Gal, Y., Hofmann, K., Whiteson, S.: Varibad: a very good method for Bayes-adaptive deep RL via meta-learning. arXiv preprint arXiv:1910.08348 (2019)
61. Xu, K., Ratner, E., Dragan, A., Levine, S., Finn, C.: Few-shot intent inference via meta-inverse reinforcement learning (2018)
62. Sohn, S., Woo, H., Choi, J., Lee, H.: Meta reinforcement learning with autonomous inference of subtask dependencies. arXiv preprint arXiv:2001.00248 (2020)
63. Bengio, Y., Deleu, T., Rahaman, N., Ke, R., Lachapelle, S., Bilaniuk, O., Goyal, A., Pal, C.: A meta-transfer objective for learning to disentangle causal mechanisms. arXiv preprint arXiv:1901.10912 (2019)
64. Huang, S., Elhoseiny, M., Elgammal, A., Yang, D.: Learning hypergraph-regularized attribute predictors. In: Proceedings of the IEEE Conference on Computer Vision and Pattern Recognition, pp. 409–417 (2015)
65. Lampert, C.H., Nickisch, H., Harmeling, S.: Attribute-based classification for zero-shot visual object categorization. IEEE Trans. Pattern Anal. Mach. Intell. **36**(3), 453–465 (2013)
66. Zhang, Z., Saligrama, V.: Zero-shot learning via semantic similarity embedding. In: Proceedings of the IEEE International Conference on Computer Vision, pp. 4166–4174 (2015)
67. Norouzi, M., Mikolov, T., Bengio, S., Singer, Y., Shlens, J., Frome, A., Corrado, G.S., Dean, J.: Zero-shot learning by convex combination of semantic embeddings. arXiv preprint arXiv:1312.5650 (2013)
68. Changpinyo, S., Chao, W.L., Gong, B., Sha, F.: Synthesized classifiers for zero-shot learning. In: Proceedings of the IEEE Conference on Computer Vision and Pattern Recognition, pp. 5327–5336 (2016)
69. Akata, Z., Perronnin, F., Harchaoui, Z., Schmid, C.: Label-embedding for image classification. IEEE Trans. Pattern Anal. Mach. Intell. **38**(7), 1425–1438 (2015)
70. Akata, Z., Reed, S., Walter, D., Lee, H., Schiele, B.: Evaluation of output embeddings for fine-grained image classification. In: Proceedings of the IEEE Conference on Computer Vision and Pattern Recognition, pp. 2927–2936 (2015)
71. Chen, Q.: Neuromorphic learning systems for supervised and unsupervised applications (2016)
72. Romera-Paredes, B., Torr, P.: An embarrassingly simple approach to zero-shot learning. In: International Conference on Machine Learning, pp. 2152–2161 (2015)
73. Antoniou, A., Storkey, A., Edwards, H.: Data augmentation generative adversarial networks. arXiv preprint arXiv:1711.04340 (2017)
74. Wang, Y.X., Girshick, R., Hebert, M., Hariharan, B.: Low-shot learning from imaginary data. In: Proceedings of the IEEE Conference on Computer Vision and Pattern Recognition, pp. 7278–7286 (2018)
75. Gao, H., Shou, Z., Zareian, A., Zhang, H., Chang, S.F.: Low-shot learning via covariance-preserving adversarial augmentation networks. In: Advances in Neural Information Processing Systems, pp. 975–985 (2018)
76. Pahde, F., Puscas, M., Wolff, J., Klein, T., Sebe, N., Nabi, M.: Low-shot learning from imaginary 3D model. In: 2019 IEEE Winter Conference on Applications of Computer Vision (WACV), pp. 978–985. IEEE (2019)
77. Hariharan, B., Girshick, R.: Low-shot visual recognition by shrinking and hallucinating features. In: Proceedings of the IEEE International Conference on Computer Vision, pp. 3018–3027 (2017)

78. Nichol, A., Achiam, J., Schulman, J.: On first-order meta-learning algorithms. arXiv preprint arXiv:1803.02999 (2018)
79. Li, Z., Zhou, F., Chen, F., Li, H.: Meta-sgd: Learning to learn quickly for few-shot learning. arXiv preprint arXiv:1707.09835 (2017)
80. Sun, Q., Liu, Y., Chua, T.S., Schiele, B.: Meta-transfer learning for few-shot learning. In: Proceedings of the IEEE Conference on Computer Vision and Pattern Recognition, pp. 403–412 (2019)
81. Snell, J., Swersky, K., Zemel, R.: Prototypical networks for few-shot learning. In: Advances in Neural Information Processing Systems, pp. 4077–4087 (2017)
82. Mukaiyama, K., Sato, I., Sugiyama, M.: LFD-Protonet: prototypical network based on local fisher discriminant analysis for few-shot learning. arXiv preprint arXiv:2006.08306 (2020)
83. Bromley, J., Guyon, I., LeCun, Y., Säckinger, E., Shah, R.: Signature verification using a "siamese" time delay neural network. In: Advances in Neural Information Processing Systems, pp. 737–744 (1994)
84. Koch, G., Zemel, R., Salakhutdinov, R.: Siamese neural networks for one-shot image recognition. In: ICML Deep Learning Workshop, vol. 2. Lille (2015)
85. Chechik, G., Sharma, V., Shalit, U., Bengio, S.: Large scale online learning of image similarity through ranking. J. Mach. Learn. Res. **11**(3), (2010)
86. Hermans, A., Beyer, L., Leibe, B.: In defense of the triplet loss for person re-identification. arXiv preprint arXiv:1703.07737 (2017)
87. Zheng, W.S., Gong, S., Xiang, T.: Transfer re-identification: From person to set-based verification. In: 2012 IEEE Conference on Computer Vision and Pattern Recognition, pp. 2650–2657. IEEE (2012)
88. Hoffer, E., Ailon, N.: Deep metric learning using triplet network. In: International Workshop on Similarity-Based Pattern Recognition, pp. 84–92. Springer (2015)
89. Vinyals, O., Blundell, C., Lillicrap, T., Wierstra, D., et al.: Matching networks for one shot learning. In: Advances in Neural Information Processing Systems, pp. 3630–3638 (2016)
90. Sung, F., Yang, Y., Zhang, L., Xiang, T., Torr, P.H., Hospedales, T.M.: Learning to compare: relation network for few-shot learning. In: Proceedings of the IEEE Conference on Computer Vision and Pattern Recognition, pp. 1199–1208 (2018)
91. Adadi, A., Berrada, M.: Peeking inside the black-box: a survey on explainable artificial intelligence (XAI). IEEE Access **6**, 52138–52160 (2018)
92. Arrieta, A.B., Díaz-Rodríguez, N., Del Ser, J., Bennetot, A., Tabik, S., Barbado, A., García, S., Gil-López, S., Molina, D., Benjamins, R., et al.: Explainable artificial intelligence (XAI): Concepts, taxonomies, opportunities and challenges toward responsible AI. Inf. Fusion **58**, 82–115 (2020)
93. Xie, N., Ras, G., van Gerven, M., Doran, D.: Explainable deep learning: a field guide for the uninitiated. arXiv preprint arXiv:2004.14545 (2020)
94. Rani, P., Liu, C., Sarkar, N., Vanman, E.: An empirical study of machine learning techniques for affect recognition in human-robot interaction. Pattern Anal. Appl. **9**(1), 58–69 (2006)
95. Wang, H.X., Fratiglioni, L., Frisoni, G.B., Viitanen, M., Winblad, B.: Smoking and the Occurence of Alzheimer's disease: cross-sectional and longitudinal data in a population-based study. Am. J. Epidemiol. **149**(7), 640–644 (1999)
96. Goodman, B., Flaxman, S.: European union regulations on algorithmic decision-making and a "right to explanation". AI magazine **38**(3), 50–57 (2017)
97. Chouldechova, A.: Fair prediction with disparate impact: a study of bias in recidivism prediction instruments. Big Data **5**(2), 153–163 (2017)

98. Hendricks, L.A., Burns, K., Saenko, K., Darrell, T., Rohrbach, A.: Women also snowboard: overcoming bias in captioning models. In: European Conference on Computer Vision, pp. 793–811. Springer (2018)
99. Bennetot, A., Laurent, J.L., Chatila, R., Díaz-Rodríguez, N.: Towards explainable neural-symbolic visual reasoning. In: NeSy Workshop IJCAI (2019)
100. Chander, A., Srinivasan, R., Chelian, S., Wang, J., Uchino, K.: Working with beliefs: Ai transparency in the enterprise. In: IUI Workshops (2018)
101. Tickle, A.B., Andrews, R., Golea, M., Diederich, J.: The truth will come to light: directions and challenges in extracting the knowledge embedded within trained artificial neural networks. IEEE Trans. Neural Networks **9**(6), 1057–1068 (1998)
102. Kim, B., Glassman, E., Johnson, B., Shah, J.: iBCM: Interactive Bayesian case model empowering humans via intuitive interaction (2015)
103. Ribeiro, M.T., Singh, S., Guestrin, C.: "why should i trust you?" explaining the predictions of any classifier. In: Proceedings of the 22nd ACM SIGKDD International Conference on Knowledge Discovery and Data Mining, pp. 1135–1144 (2016)
104. Nguyen, A., Yosinski, J., Clune, J.: Deep neural networks are easily fooled: high confidence predictions for unrecognizable images. In: Proceedings of the IEEE Conference on Computer Vision and Pattern Recognition, pp. 427–436 (2015)
105. Moosavi-Dezfooli, S.M., Fawzi, A., Frossard, P.: Deepfool: a simple and accurate method to fool deep neural networks. In: Proceedings of the IEEE Conference on Computer Vision and Pattern Recognition, pp. 2574–2582 (2016)
106. Yuan, X., He, P., Zhu, Q., Li, X.: Adversarial examples: attacks and defenses for deep learning. IEEE Trans. Neural Networks Learn. Syst. **30**(9), 2805–2824 (2019)
107. Law, J.: Robust statistics-the approach based on influence functions. J. Roy. Stat. Soc. Ser. D (The Statistician) **35**(5), 565–566 (1986)
108. Basu, S., Kumbier, K., Brown, J.B., Yu, B.: Iterative random forests to discover predictive and stable high-order interactions. Proc. Natl. Acad. Sci. **115**(8), 1943–1948 (2018)
109. Zeiler, M.D., Krishnan, D., Taylor, G.W., Fergus, R.: Deconvolutional networks. In: 2010 IEEE Computer Society Conference on Computer Vision and Pattern Recognition, pp. 2528–2535. IEEE (2010)
110. Selvaraju, R.R., Cogswell, M., Das, A., Vedantam, R., Parikh, D., Batra, D.: Grad-cam: visual explanations from deep networks via gradient-based localization. In: Proceedings of the IEEE international conference on computer vision, pp. 618–626 (2017)
111. Zhou, B., Khosla, A., Lapedriza, A., Oliva, A., Torralba, A.: Learning deep features for discriminative localization. In: Proceedings of the IEEE Conference on Computer Vision and Pattern Recognition, pp. 2921–2929 (2016)
112. Blunsom, P., Cho, K., Cohen, S.B., Grefenstette, E., Hermann, K.M., Rimell, L., Weston, J., Yih, W.T.: Proceedings of the 1st Workshop on Representation Learning for NLP. In: Proceedings of the 1st Workshop on Representation Learning for NLP (2016)
113. Arras, L., Horn, F., Montavon, G., Müller, K.R., Samek, W.: " what is relevant in a text document?": an interpretable machine learning approach. PLoS ONE **12**(8), e0181,142 (2017)

114. Shrikumar, A., Greenside, P., Kundaje, A.: Learning important features through propagating activation differences. arXiv preprint arXiv:1704.02685 (2017)
115. Zintgraf, L.M., Cohen, T.S., Adel, T., Welling, M.: Visualizing deep neural network decisions: prediction difference analysis. arXiv preprint arXiv:1702.04595 (2017)

Obstacle Awareness Subsystem for Higher Exoskeleton Safety

Pavel Blažek[1], Josef Bydžovský[1], Robert Griffin[2], Karel Mls[1](✉), and Brandon Peterson[2]

[1] University of Hradec Králové, Hradec Králové, Czech Republic
{pavel.blazek,josef.bydzovsky,karel.mls}@uhk.cz
[2] Florida Institute for Human and Machine Cognition, Pensacola, FL, US
{rgriffin,bpeterson}@ihmc.us

Abstract. Exoskeletons are electromechanical systems whose main purpose is to facilitate the workload or to enable the mobility of their users. They combine human intelligence and machine power to strengthen both - thus adding machine power to people and giving machines human intelligence. Unlike robots, the exoskeleton is controlled by a human operator in an architecture called man-in-the-loop. In this architecture, the human operator not only issues commands to the exoskeleton but also receives feedback from it. This feedback can come from the exoskeleton itself or the environment. Besides, exoskeletons can assist a person in making decisions by performing complex calculations and other types of data manipulation in real-time.

This chapter introduces an example of the solution of the environmental awareness subsystem of the exoskeleton design for people with lower-body disabilities, including its design, implementation and results of practical tests.

Keywords: Exoskeleton · Environment awareness · Man-in-the-loop

1 Introduction

In the middle of the 20th century, scientists realized the limited possibilities of artificial intelligence at that time for its use in robotics. Therefore, it was not until the end of the same century that some researchers suggested solving this problem through the involvement of human intelligence. They gave rise to the concept of so-called "humachine" systems and defined a specific method of their construction [1]. A typical representative of this concept - an exoskeletal system - is a structure that creates a pair of man and machine and sets up their relationship in an intelligent human-centred system.

The historical development of exoskeletons can be divided into three main phases. In the first phase, exoskeletal systems were usually one of three types. The first type represents a remote-controlled arm for telerobotics [2, 3].

These systems can be used in situations that are dangerous to humans, such as chemical disasters, bomb deactivation or work in the space. These systems can also be

J. Paralič et al. (Eds.): DISA 2020, AISC 1281, pp. 59–71, 2021.
https://doi.org/10.1007/978-3-030-63872-6_3

used in situations that require extraordinary strength or precision that one is not capable of. Initially, these systems had to be connected to a stationary structure for safe use. Only later, when these exoskeletons became light enough, they could be attached directly to the user himself.

The second type of exoskeletons in the first developmental phase are systems for accurate measurement of the position of individual fingers [4, 5]. These systems can be used as an alternative interface to other motion digitizing devices for the film or gaming industry.

The third type are rehabilitation devices, such as gloves for people with limited hand control [6], or orthosis medical aids for people with disabilities similar to those that will be the subject of this chapter.

The second development phase brought force feedback to exoskeleton systems using haptic technology [7] or force reflection [8]. These technologies give the user a sense of touch that helps with the operation of the exoskeleton. Controlling the exoskeleton system with this function is more like using an extension of our body instead of controlling something with which we have no complete connection.

The current development of the third phase is focused on new exoskeletons with very specific purposes and high integration of new technologies. Along with force feedback, this high integration also includes data fusion and data transformation, which provides the user with useful data and helps him make decisions based on calculations that would take him a long time. Other new technologies include three degrees of joint freedom, micro actuators, Hall sensors, and motion tracking.

Exoskeletons have great potential to change our lives, whether by improving our capabilities or restoring our abilities after a serious injury, whether in industry, the military or healthcare. Nevertheless, they still have a long way to go and many problems will need to be solved before they become a regular part of our daily lives. One of these problems, and probably the most important, is to ensure the safety of their operators and other people around them. Exoskeletons must be able to protect their operators not only from the negative consequences of interacting with the environment but also from their own mistakes.

This chapter introduces the solution of the environmental awareness subsystem of the exoskeleton intended for people with lower-body disabilities, including its design, implementation and the results of practical tests.

2 Exoskeletons

Exoskeletons are mechanical devices designed to help or improve the abilities of their users. They combine human intelligence and machine power to strengthen both - to strengthen human power and machine intelligence. Ultimately, this symbiosis is designed to serve a person and facilitate or enable him to fulfil his tasks. Exoskeletons became popular in the late 20th century in highly developed countries such as the United States, Japan and Germany. This has been made possible by developments in mechanical, electrical and software engineering and the biological and materials sciences. They can be used in many areas such as industry, military, healthcare, entertainment and others [9].

Unlike robotic systems, where the whole system is controlled autonomously - for example by artificial intelligence or utilizing a deterministic program - the exoskeleton is controlled in part or in full by a human operator. We say that exoskeletons use a man-in-the-loop architecture. In this architecture, the human operator not only controls the exoskeleton but also receives feedback from it. This creates a loop in which the operator enters exoskeleton commands, the exoskeleton executes them and gives the operator feedback. The information in this feedback can come either from the exoskeleton itself or from the affected environment. Besides, exoskeletons can help the operator with decision-making by performing complex calculations, data fusion, and/or other types of data processing that would be difficult or impossible for the human operator in real-time.

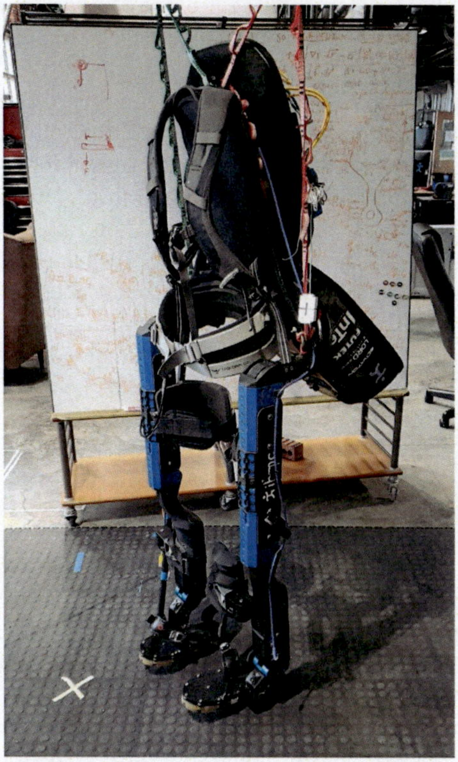

Fig. 1. Mina v2 exoskeleton from the Institute for Human & Machine Cognition (IHMC)

Mina v2 (Fig. 1) is an exoskeleton designed for people with lower body motor problems. Its goal is to allow them to walk again with the support of crutches. This is the third successive exoskeleton created by a group of robotics, exoskeletons and human robotic interdependence on IHMC after Mina v1 [10] and NASA X1 exoskeleton [11]. Its design is also influenced by the development of the Hopper exoskeleton, primarily intended as an exercise device for use by astronauts during their long-term space missions [12].

The mechanical construction of Mina is based on an original design using an assembly composed of carbon parts. It includes six electric drives built into this structure - two for hip movement, two for knee movement and two for ankle movement. Fastening straps are located on various parts of the structure to fix the pilot's legs and body to the exoskeleton, which ensure the pilot's safety during movement. Most of the control electronics, including the battery, is housed in a backpack. The whole exoskeleton weighs approximately 34 kg. Mina v2 is designed specifically for one particular pilot and should not be used by anyone with different body proportions. In its current form, it serves as a prototype for future universal versions, which will be adjustable for different pilots.

The actuators are custom linear linkage actuators. They have been designed in the group of robotics, exoskeletons and human robotic interdependencies and include features such as a frameless electric motor, integrated electronics, an integrated motor amplifier and a controller for distributed control at a common level. Thanks to the Allied Motion HS02303 motors, they can produce a torque of 110 Nm and a speed of 9 rad/s, without load. These motors operate at the peak current of 20 A.

For computations, Mina v2 uses COM Express Type 6 computer from ADLINK Technology Inc. With a custom Ubuntu kernel. The controlling software is written in Java using the POSIX real-time threads. EtherCAT is used for communication between the computer and the actuators. The battery, used to power the exoskeleton, is 48V, 480Wh lithium-ion battery for electric bicycles. When fully charged, this battery can supply power to most of the components for 2.5 h of regular use.

The pilot controls the Mina v2 exoskeleton via the Raspberry Pi3 microcomputer with a screen, a thumb joystick, and an index finger button, all mounted on the right crutch. The entire interface runs on its own battery and communicates with the exoskeleton on the TPC/IP over the Ethernet cable or Wi-Fi. The pilot chooses what action he wants to take – such as the whole step forward or sat down, and performs it by pressing the index finger button.

3 Obstacle Detection

Upright walking is one of our daily activities that do not require too much effort. Thanks to our nervous system this process is highly automated and does not take all our attention. However, this does not apply to pilots of exoskeletons with lower body impairments. For them, walking requires their full attention and physical effort. The following subchapters will describe this problem in more detail and the reasons supporting the implementation of an obstacle awareness system that should solve this problem will be discussed.

People who can walk independently have full control over this process. They can control the muscles of the legs (actuators) and also receive the feeling of touch and information about the position of their limbs (sensors). Sight is also an important sense, giving the nervous system visual information about the path ahead. Exoskeletons like Mina and other similar exoskeletons are intended for people who are not able to control the muscles and have no feeling in the legs, but not for people with visual impairments. This may raise the question of why we should create a vision-like exoskeleton sensor subsystem if it is a fully functional "subsystem" that the pilot can provide.

The problem is how to integrate visual perceptions with muscle control and tactile sensation. In people without lower body involvement, this integration occurs in the central nervous system, especially in the brain [13]. This integration is very important for many reasons, such as automating movement, maintaining balance, making quick decisions or preventing tripping. Unfortunately, this type of integration is not possible in the brain of an exoskeleton operator. In this case, we have to work with two different integration centres - the operator's brain and the exoskeleton software. This raises several issues that need to be addressed to increase the synergy and practicality of the whole facility.

3.1 Design

How and whether the exoskeleton will move is decided by its pilot. Therefore, if there is an obstacle on the way, we can simply leave the decision to stop and change direction to the pilot. Nevertheless, we have learned from history that the human factor is the main cause of problems in modern systems, which have a higher level of automation, but some decisions are still dependent on users [14]. The obstacle awareness system was therefore also designed as a safety subsystem to ensure the safety of the pilot.

Through this subsystem, the pilot will be informed of the detected obstacles and of the exoskeleton's decision to prevent further forward movement if an obstacle is dangerously close. This will provide feedback to the pilot, which will increase synergy and ensure that the pilot feels more comfortable and knows what is happening and why. For this purpose, we can theoretically use any of the five basic senses: sight, hearing, smell, taste and touch. Of course, in case of touch, we cannot use those parts of the body that are affected by the pilot's disability.

3.2 Implementation

The Mina v2 exoskeleton described above was used for the practical implementation of the obstacle awareness subsystem. One of the main requirements for the development of this subsystem was the ability to easily and quickly remove a prototype from the exoskeleton if needed. On the other hand, there was a requirement that the possible integration of the resulting solution into the exoskeleton be as simple as possible. For this reason, additional hardware and software components were used to develop the subsystem. These components and their functions will be described in the following subsections.

Up^2.

To meet the above requirements, a separate microcomputer was used for the LIDAR module, detection and evaluation algorithms and ROS wrapper instead of a central exoskeleton computer. This made it possible to quickly remove the obstacle awareness system from the exoskeleton, if necessary, and also facilitated its development, as it was not necessary to switch on the entire exoskeleton to test the obstacle awareness system.

Up^2 was chosen as the ideal microcomputer for development. This 32-bit microcomputer is manufactured by Intel Corporation and is similar to other microcomputers such as Raspberry Pi or Odroid-xu4. It can have different Intel processors, such as Celeron N3350, or Pentium N4200, operation memory up to 8 GB, storage capacity up to 128 GB, and other customized components. It has many ports including USB 3.0, two gigabit Ethernet ports, HDMI, and 40 pins general-purpose bus. It uses 5V DC in the 4–6 A range for power supply. It supports operating systems such as Linux and Android, but it is also able to run the desktop version of Microsoft Windows 10 operating system. The server version of Ubuntu 14.04 was chosen to work in an environment similar to the exoskeleton itself.

Hololens and Unity.
Simple visualization of additional information in Microsoft HoloLens did not require the use of complicated technology. There were only basic requirements - fast, simple and wireless communication with Up^2 to obtain information about obstacles, and a clear presentation of this information on holographic displays. For this purpose, Unity was chosen as a suitable environment [15].

Unity is a cross-platform game engine developed by Unity Technologies. It is very popular for development - since 2018 50% of mobile games and 60% of augmented and virtual reality projects have been developed in it. It uses C # as a scripting language and supports more than 25 platforms such as Windows, Android, iOS, WebGL, PlayStation 4, Xbox One, Oculus Rift, SteamVR, Facebook Gameroom and more. Unity has also been used in the film industry as well as for artificial intelligence training (DeepMind and Alphabet Inc.).

Unity supports two-dimensional augmented reality and supports Microsoft HoloLens and its operating system, including access to wireless communication. This allows the use of Wi-Fi and the establishment of one-way UDP communication from Up^2 to Microsoft HoloLens. The Wi-Fi router was an integral part of this network because neither the USB Wi-Fi stick for UP^2 nor the Microsoft HoloLens Wi-Fi adapter can function as a router or operate in point-to-point mode.

Depth Camera.
Another issue that needed to be addressed was how and where the depth camera (or depth cameras) should be mounted on the exoskeleton. After evaluating the depth camera parameters, such as size, USB port position, or viewing angle and mounting options on the exoskeleton itself, it was decided to create a 3D printed depth camera bracket that can be easily attached to the outside of the exoskeleton thigh. Using only one depth camera on one thigh would cause a viewing angle problem and it would not be possible to cover a sufficiently wide area in front of the exoskeleton. Unfortunately, the exoskeleton has no central location ideal for connecting a single depth camera. Figure 2 shows the design of the 3D holder and its position on the left thigh.

ROS and ROS Wrapper.
One of the requirements for the obstacle awareness system was the possible integration into the Mina v2 control software. For this reason, the ROS wrapper was chosen as one of the ways to access the above-mentioned depth camera. The ROS wrapper is a bridge between many Intel RealSense and ROS products [16].

ROS stands for Robotic Operating System. It is an open-source framework for network communication between heterogeneous systems [17], so its use is not limited to robotic applications only as the name might suggest. It combines server-client and peer-to-peer approaches in the publishing-subscribing mechanism. Each node in the network can act as a publisher and/or as a subscriber. The publisher waits for two events, the first one is when a subscriber subscribes for a topic that the publisher publishes. The

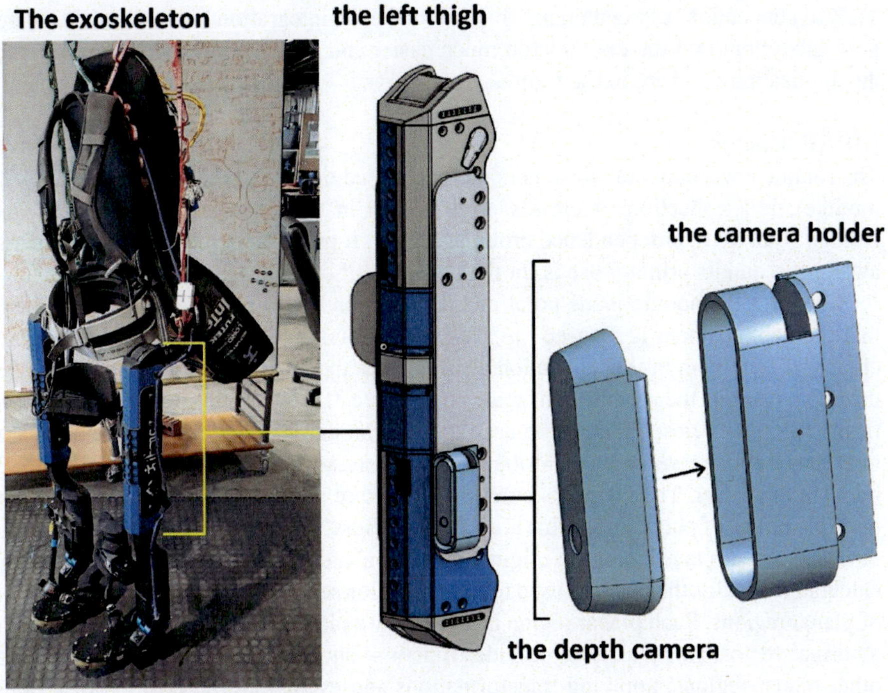

Fig. 2. Location of the depth camera on the Mina v2 exoskeleton.

second one is when the publisher receives a message, that is supposed to be published. When this occurs, the publisher sends this message over the network to all subscribers that have previously subscribed to the topic. Moreover, each node can connect to and disconnect from the network at any time during runtime without causing an error on any other node.

ROS has implementations in many languages, such as Python, C++, Lisp, and more. In addition to the network communication described above, it also provides features such as hardware abstraction, low-level device control, implementation of commonly-used functionality in robotics, various styles of the network communication, and package management. One of the main features of ROS is that it is easy to integrate into existing software but also easy to remove. Scalability is another important aspect of ROS and it includes unit and integration tests.

To share data between stand-alone applications, we need a communication channel between the obstacle awareness subsystem and the exoskeletons controlling software. The simplest solution was to use the IHMC's ROS2 implementation in Java that Mina already uses [18]. ROS2 system is very similar to the original ROS - it is built on similar principles, but with additional tools, with a wider range of supported devices and wit no compatibility issues with the previous version. An Ethernet cable was used for the physical connection between Up^2 and the exoskeleton. The advantage of ROS2 is that both end elements that communicate with each other can run in the same program, in different programs, but on the same computing unit, or different computing units with

TCP/IP connection between them. This will make the integration of the obstacle awareness subsystem into the exoskeleton much easier and eliminate the need to reprogram the application software in the future.

LIDAR Module.
The robotic environmental awareness module, based on the LIDAR sensor (the LIDAR module), is a collection of classes programmed in the Robotics, Exoskeletons and Human Robotic Interdependence group at IHMC. It provides classes of environmental awareness, and its primary use is for the detection of obstacles and stairs [19].

The LIDAR module loads point cloud collection that represents the environment, that the depth camera is focused on. This point cloud is then cached with other point clouds. If each item in this collection also includes the position and the view vector of the depth camera, the module can create an accurate 3D representation of the real world from multiple entries. This can be used to 3D scan an object from multiple angles or to create a 360° representation of the environment. An octree is then created from all items in the buffer. The Octree is a tree-like structure where each node is composed of multiple points of point clouds that are relatively close to each other [20]. This grouping of nodes reduces the size of the original collection and allows the normal vectors to be calculated. Finally, the octree is used for planar region segmentation, creating a collection of planar regions. Each planar region is defined by a plane, a bounding concave hull and a normal vector. This class also provides functions such as detecting intersections with other planar regions, applying transformations and evaluating the position of a point relative to a planar region.

Both processes are parameterized so that the user can adapt the use of this model to the specific case. The module is primarily designed to work with data from LIDAR sensors, but it can also process stereo vision point cloud or any other form of a point cloud. An important component of this module are classes that support the creation and use of datasets. Last but not least, it includes a graphical interface that helps to visualize the entire scene.

4 Results

Due to the flow of information in the obstacle awareness system, most of its parts cannot be tested separately, but only in combination with others. The following paragraphs will show the results by the component of the entire system.

4.1 Testing the Depth Camera

The first thing for testing was the depth camera, more specifically its precision. Because, in the obstacle awareness system, we need the maximum precision we can get, it is important to know the limits of the camera itself.

There are a lot of different settings that allow its user to tweak the depth cameras results in the slightest detail of its functioning. There is a possibility to use a calibration software that creates and tweaks settings based on the specific environment. This calibration tool was used to create settings in the testing environment of the obstacle awareness system and these settings were used throughout the whole testing process.

Except for settings, an environment itself plays a very important factor that affects the quality of output. Because the D435 depth camera is sensitive to light, the illumination of the scene is very important [21]. Generally, insufficient light in an environment will negatively affect precision or even does not allow any kind of detection at all. During testing, we found out that this threshold in the D435 camera is higher compared to the threshold in the human eye. That means that the depth camera needs more light for proper detection than a human eye. The opposite extreme is too much light which also does not allow proper functioning. This problem is especially problematic on shiny surfaces where the illumination of the surface can be very strong. Even here we found out that a human eye surpasses capabilities of the depth camera in the matter of tolerance.

4.2 Testing the LIDAR Module

Another part tested was the LIDAR module and its quality. This module is stochastic which means that we can get different outputs for the same input. In our case, the input comes from the depth camera that already has an error in its output, for this reason, the test was conducted on two datasets. Both datasets were created in the same scene simulating a staircase with the only difference in the horizontal distance between the depth camera and the staircase. In the first dataset, the horizontal distance was approximately 0.52 m, in the second dataset, the horizontal distance was approximately 1.37 m. Figure 3 shows an example of output from datasets as they were processed by the LIDAR module.

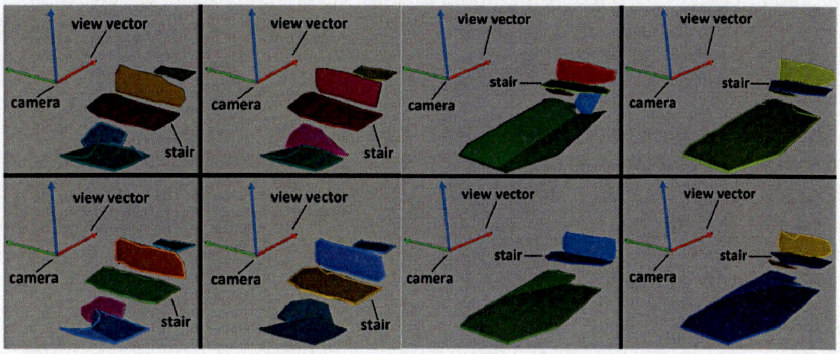

Fig. 3. Visual outputs from the LIDAR module for two datasets.

Just by visual examination, it can be seen that the differences between results within the same dataset are bigger in the second dataset. We can observe that in the first dataset we have the same number of planes in all four runs, but in the second dataset, the number of planes varies. Another thing we can observe is the differences in shapes of the planes that are located in the same place in all sample pictures of that particular dataset.

To determine the significance of these differences mathematically, we picked two variables that are important for the detection and evaluation algorithms. The first one is the angle between the view vector of the camera and the normal vector of the plane. The second variable is the horizontal distance between the camera and the plane representing the stair. In both cases, the range between the maximum and the minimum values are within one degree. This shows the high precision of the LIDAR module. This information can be used for better determination of the boundaries within the angle between the camera's view vector and the plane's normal vectors to be considered as a possible step.

Numerical evaluation of the difference between the maximum and the minimum values of measured distances is less than 2 cm in both datasets. This is sufficient precision for the detection and evaluation algorithms. This also does not show any major difference between these two datasets and their horizontal distance differences.

Based on evaluations of the angles and the horizontal distances we can deduce, that the differences in the visual representation mentioned above are not significant in the matter of precision in the calculations of the LIDAR module.

4.3 Testing the Up^2

An important aspect of the whole obstacle awareness subsystem was its capability to operate in real-time. Because the information flow is almost chain-like, parallelism could not be used. On the other hand, this chain should be processed in a cycle as many times as possible. It follows that only the slowest operation is the limiting factor of the whole chain.

The depth camera can operate at the speed up to 90 fps. The setting in our experiments was around 35 fps which was still sufficient. The only tasks of the Microsoft HoloLens were to receive the information over the Wi-Fi in the form of the UDP packets and print the information on the holographic display. Its computational capacities are more than sufficient for these tasks to be executed in real-time. The exoskeleton itself has enough computational power needed for its own calculations plus reserves, so it is not the weakest part neither.

The remaining four parts (the ROS wrapper, the ROS bridge, the LIDAR module, and the detection and evaluation algorithm) conduct their calculations on the Up^2 microcomputer. These calculations can be very demanding and put a lot of pressure on the Up^2. Therefore, this test was conducted to determine, if the Up^2 is the weakest part in the information flow and if so, what is the maximum speed at which the whole system can operate.

The ROS wrapper on its own does not take much of the CPU time. Its only purpose is to receive information from the depth camera and provide it to the ROS bridge in the form of a publisher-subscriber relationship, therefore, the complexity is C where C is a constant. The ROS bridges receive this data and transform them into a form that the LIDAR module can use. This transformation is conducted on each point from a point cloud independently on the others therefore the complexity is N where N is the number of points.

The complexity of the LIDAR module is approximately quadratic. The exact complexity is almost impossible to determine because its implementation contains a lot of branching in code which can dramatically change its complexity. Also, which point from the point cloud will use which branch depends on many aspects like already processed points, the overall structure of the scene, an order of the points, etc. Also, the fact, that the LIDAR module is stochastic would make the complexity determination very difficult.

This fact was demonstrated in the test, in which the LIDAR module conducted 500 calculations on each of the 9 different datasets. These datasets contained point clouds with a specific number of points ranging from 40,000 to 120,000 points. During this test, the time needed for these calculations was measured to determine the average time for each dataset. This test was conducted on the Up^2 microcomputer that was part of this system. As was expected, with the increasing number of points, the calculation time is also increasing. The quadratic complexity means that the calculations of the LIDAR module on the Up^2 are the weakest part of the obstacle awareness system because of the time they need and also because of the wide range of how long these calculations can take.

The LIDAR module running on the Up^2 microcomputer is not capable to provide output in real-time. This problem should be eliminated when the exoskeleton's computer will be used for calculations instead of the Up^2. For testing purposes, this problem was alleviated by two mechanisms. The first one was using the depth camera's ability to set the maximum depth. Every point beyond this maximum depth was excluded from the point cloud and only points close enough were used for additional processing. In our case, the maximum distance was reduced from the default 10 m to 3 m.

The second mechanism was to restrict the area in which the LIDAR module searches for planes. This feature is called a bounding box and is defined by two 3-dimensional points that represent a cuboid. Only points from the point cloud that are inside this cuboid can be considered to be a part of planes, all other points are ignored. Both these measures reduced the number of points and therefore the average time needed for calculations was also reduced.

5 Conclusion

The subject of interest of this chapter was the design of the obstacle awareness subsystem for the exoskeleton Mina v 2, developed in the department of Robotics, Exoskeletons, & Human Robotic Interdependence at IHMC, Florida. Initially, a debate with members of the exoskeleton team and members of other teams were conducted. Together, a solution specifically for the case using a combination of already well-tested technologies were suggested. During the following implementation, several practical issues were encountered that were solved "on the run". When the whole system was completed and all sub-technologies worked together as expected, development of the detection and evaluation algorithms followed. This led to three algorithms each with a specific purpose and introducing several changes in the used technologies. Tests were conducted in the physical testing environment simulating a staircase going up during which outputs of the algorithms were compared with measurements from this environment. As a result, functionality of a system that warns the pilot of the exoskeleton and the exoskeleton itself about stairs in its path and provides additional information useful for safe movement were verified.

The obstacle awareness subsystem, as described in this chapter, is ready to be used in similar situations in which it was tested. However, that does not mean that its development will stop there. On the contrary, several areas can be improved and others can be newly implemented to improve the whole system. Some of them are described in the following paragraphs.

The next step that should be implemented as soon as possible is the integration of the whole obstacle awareness subsystem into the exoskeleton. That means using the exoskeleton's computer instead of Up^2. This will eliminate most issues with the system's ability to operate in real-time that was addressed in this chapter. This will also save battery capacity and space in the exoskeleton's backpack, where the Up^2 was placed for testing purposes. The exoskeleton's main computer has all the necessary peripherals such as USB ports for the depth cameras and USB Wi-Fi stick. On the software level, the control software of the exoskeleton is written in Java language, therefore it will be easy to create an additional threat for the obstacle awareness subsystem and integrate it to the exoskeleton's control software.

Another way how to improve the real-time operability will be switching from point clouds to depth images. A depth image is a 2-dimensional image where, instead of the colour information, each pixel carries depth information. This data structure should be easier to use than a point cloud which would decrease the time needed for plane detection and increase the operation speed of the whole system. This would also mean using different technology instead of the LIDAR module that works with depth images instead of point clouds in a similar way. That means planes detection, their normal vectors calculations, and recalculations into X, Y, and Z values corresponding to the physical world. This would be a huge change that might even require a change in the obstacle and detection algorithms themselves.

For the pilot's comfort, big and relatively heavy Microsoft HoloLens can be replaced with a smaller and lighter device that can provide similar functionality. Ideal candidates are smart glasses like Focals 2.0 from North Inc. or Vuzix Blade smart glasses from Vizux corporation. From all the features that Microsoft HoloLens offers only holographic display for 2D rendering and Wi-Fi adapter were used. Both smart glasses mentioned above have the same or similar features.

Because none of the tests of the obstacle and awareness systems was conducted in the situation of going the staircase down, testing this scenario should follow. This also requires changes in the detection and evaluation algorithms for this special case. Currently, the third algorithm just detects a closes stair (both up and down) and return its distance towards the exoskeleton. Not yet addressed issue is that when going the stair down the camera is unable to "see" the beginning (end) of this stair because it is hidden behind current stair or floor. A new safety measure that might be implemented to increase the pilot's safety is the calculation of the surface of the following stair to make sure, that the step is wide and long enough. This task should be simple considering the provided information from the LIDAR module.

References

1. Canjun, Y., Ying, C., Yongxiang, L.: Study on the humachine intelligent system and its application. Chin. J. Mech. Eng. **36**(6), 42–47 (2000)

2. Yang, C.J., Zhang, J.F., Chen, Y., Dong, Y.M., Zhang, Y.: A review of exoskeleton-type systems and their key technologies. Proc. Inst. Mech. Eng. Part C J. Mech. Eng. Sci. **222**(8), 1599–1612 (2008)
3. Repperger, D. W., Remis, S. J., Merrill, G.: Performance measures of teleoperation using an exoskeleton device. In: Proceedings of IEEE International Conference on Robotics and Automation, pp. 552–557. IEEE. (May 1990)
4. Jau, B. M.: Anthropomorphic exoskeleton dual arm/hand telerobot controller. In: IEEE International Workshop on Intelligent Robots, pp. 715–718. IEEE. (October 1988)
5. Wright, A.K., Stanisic, M.M.: Kinematic mapping between the EXOS Handmaster exoskeleton and the Utah/MIT dextrous hand. In 1990 IEEE International Conference on Systems Engineering, pp. 101–104. IEEE. (August 1990).
6. Brown, P., Jones, D., Singh, S.K., Rosen, J.M.: The exoskeleton glove for control of paralyzed hands. In: Proceedings IEEE International Conference on Robotics and Automation, pp. 642–647. IEEE. (May 1993).
7. Okamura, A.M.: Methods for haptic feedback in teleoperated robot-assisted surgery. Ind. Robot. **31**(6), 499–508 (2004)
8. Kim, Y.S., Lee, J., Lee, S., Kim, M.: A force reflected exoskeleton-type masterarm for human-robot interaction. IEEE Trans. Syst. Man Cybern. Part A Syst. Hum. **35**(2), 198–212 (2005)
9. Bai, S., Virk, G.S., Sugar, T.: Wearable exoskeleton systems: Design, control and applications. Institution of Engineering and Technology, London (2018)
10. Neuhaus, P. D., Noorden, J. H., Craig, T. J., Torres, T., Kirschbaum, J., Pratt, J. E.: Design and evaluation of mina: a robotic orthosis for paraplegics. In: 2011 IEEE international conference on rehabilitation robotics, pp. 1–8. IEEE. (June 2011).
11. Rea, R., Beck, C., Rovekamp, R., Neuhaus, P., Diftler, M.: X1: a robotic exoskeleton for in-space countermeasures and dynamometry. In: AIAA Space 2013 Conference and Exposition, pp. 1–8. (2013).
12. Beck, C.E., Rovekamp, R.N., Neuhaus, P.D.: The Hopper: a wearable robotic device testbed for micro-gravity bone-loading proof-of-concept. In: Human Research Program Investigators' Workshop, 13 January 2015–15 January 2015, Galveston, TX, United States, p. 1. (2015)
13. Inman, V.T., Ralston, H.J., Todd, F.: Human walking. Williams & Wilkins, Baltimore (1981)
14. Parasuraman, R., Riley, V.: Humans and automation: use, misuse, disuse, abuse. Hum. Factors **39**(2), 230–253 (1997)
15. Evans, G., Miller, J., Pena, M.I., MacAllister, A., Winer, E.: Evaluating the Microsoft HoloLens through an augmented reality assembly application. In: Proceedings of SPIE 10197, Degraded Environments: Sensing, Processing, and Display 2017, 101970V, p. 16 (May 2017)
16. Mishra, R., Javed, A.: ROS based service robot platform. In: 2018 4th International Conference on Control, Automation and Robotics (ICCAR), pp. 55–59. IEEE. (April 2018).
17. Quigley, M., Conley, K., Gerkey, B., Faust, J., Foote, T., Leibs, J., Ng, A.Y.: ROS: an open-source Robot Operating System. In: ICRA workshop on open source software, vol. 3, No. 3.2, p. 5. (May 2009).
18. Maruyama, Y., Kato, S., Azumi, T.: Exploring the performance of ROS2. In: Proceedings of the 13th International Conference on Embedded Software, p. 10 (October 2016).
19. Himmelsbach, M., Mueller, A., Lüttel, T., Wünsche, H.J.: LIDAR-based 3D object perception. In: Proceedings of 1st international workshop on cognition for technical systems, pp. 1–7 (October 2008).
20. Hornung, A., Wurm, K.M., Bennewitz, M., Stachniss, C., Burgard, W.: OctoMap: an efficient probabilistic 3D mapping framework based on octrees. Auton. Robot. **34**(3), 189–206 (2013)
21. Ahn, M.S., Chae, H., Noh, D., Nam, H., Hong, D.: Analysis and noise modeling of the Intel RealSense D435 for mobile robots. In: 2019 16th International Conference on Ubiquitous Robots (UR), pp. 707–711. IEEE. (June 2019).

Intelligent Analysis of Big Data

Knowledge-Based Approaches to Intelligent Data Analysis

Peter Bednár(✉), Ján Paralič, František Babič, and Martin Sarnovský

Department of Cybernetics and Artificial Intelligence, Technical University of Košice, Letná 9, 04200 Košice, Slovakia
{peter.bednar,jan.paralic,frantisek.babic, martin.sarnovsky}@tuke.sk

Abstract. One of the most fundamental phenomena heavily influencing the digital society is Big Data. It is crucial not only to collect and analyze vast amounts of data but do it in an intelligent way. We believe that in order to do so, there needs to be a suitable interplay between the knowledge already known in the given application domain (background knowledge) and the knowledge inductively gained from data utilizing various data analysis techniques. We call it a knowledge-based approach to data analysis or intelligent data analysis. In this chapter, we will focus on two main types of the knowledge-based approach to data analysis. We start with the introduction of the semantic modelling of data analytics processes, which can efficiently cover an explicit form of background knowledge. The main focus here will be on the conceptualization of domain knowledge shared between the domain expert and data scientist and modelling of data mining workflows in order to achieve reproducibility and reusability. The second situation is typical for medical application, where the prevalent amount of background knowledge tends to stay tacit. In such a situation, the human-in-the-loop approach is a way how to perform data analysis intelligently. For both of these types of knowledge-based data analysis, specific case studies are presented to show how intelligent data analysis works in practice.

Keywords: Data analysis · Big data · Knowledge-based approach

1 Introduction

Recently, important breakthroughs in science or industry were achieved by applying data analytics methods on large datasets. The data analysis process is interactive and iterative and requires cooperation between the data scientists and domain experts. Additionally, implementation of the data analysis processes requires complex infrastructure with many technologies and software/hardware dependencies, which can have an impact on the final results of the analysis. The complexity of the data analysis processes can be tackled by means of knowledge-based approaches.

Knowledge in the context of data analysis has an important role both, at the input of the data analytical process, often called background knowledge, which is, e.g. knowledge about the application domain, but also the knowledge about the data mining process

J. Paralič et al. (Eds.): DISA 2020, AISC 1281, pp. 75–97, 2021.
https://doi.org/10.1007/978-3-030-63872-6_4

itself, both from the procedural as well as the technical side. This knowledge may have an explicit form (procedural and technical aspects and partially also about application domain) as well as tacit form (application domain knowledge). Knowledge at the output of the data analytical process is expected to be valid, new and potentially useful. It predominantly has an explicit form of, e.g. a suitable machine learning model. In this chapter, we focus on the input part of the data analytical processes, where knowledge can have both dominantly explicit form in applications like process industry or network intrusion detection. However, there are application domains, like, e.g. medical applications, where input knowledge is dominantly of tacit nature. Whereas for explicit knowledge semantic modelling approach may help to cope with the complexity problem, for tacit knowledge a suitable involvement of domain expert in particular phases of the data analytical process is needed (so-called human-in-the-loop concept).

Semantic modelling of particular aspects of data analytical processes is a crucial aspect of intelligent data analytics from two main reasons. Firstly, it enables to (at least partially) automatize the analytic processes of rapidly increasing amounts of data available in growing number of application domains. Secondly, it makes it easier for people to set the fundamental constraints and requirements on data analysis tasks and their results as well as understand the knowledge discovered in such tasks. Both reasons lead to increased intelligence of (big) data analytics achieving better efficiency of data analytic processes on one hand side and better understandability and increased trust in knowledge discovered by these data analytic processes on the other hand side.

The rest of the chapter is structured as follows. The second section provides an introduction and state of the art in data analytics and semantic modelling of data analytical processes. In the third section, we present an original semantic framework for modelling of data analytical processes. Section four presents two particular case studies where our semantic framework has successfully been applied.

Section five presents the human-in-the-loop approach first methodologically and then on the selected medical experimental scenarios implemented in strong cooperation with domain experts to present the second type of knowledge-based approach to data analysis. Chapter six concludes with a summary of the main findings and suggestions for open research problems in the intelligent data analytics domain.

2 Semantic Modelling of Data Analytics Processes

2.1 The Data Analytics Processes

The process of data analytics can be described by the standard methodologies such as Cross-Industry Standard Process for Data Mining (CRISP-DM) [1] or Sample, Explore, Modify, Model, and Assess (SEMMA), which breaks the process of data analysis into six major steps (see Fig. 1):

Problem (Business) Understanding. This initial phase focuses on understanding the objectives and requirements of the data analysis from a business perspective, i.e. how the production process can be optimized by data analytics methods. During this phase, the domain experts in cooperation with the data scientists analyze the overall production process and its steps (production segments) and specify which key-performance indicators will be optimized. This knowledge is then converted into the data analysis problem definition (e.g. classification, regression, anomaly detection, etc.) by data scientists.

Data Understanding. The data understanding phase starts with an initial data collection and proceeds with activities in order to get familiar with the data, to identify data quality problems, to discover first insights into the data, or to detect interesting subsets to form hypotheses for hidden information. During this phase, data scientists with the cooperation of domain experts will select and describe the subset of relevant data required to achieve optimization objectives specified in the phase of Problem understanding. The part of a data description is also the identification of the already known dependencies between the data and Key Performance Indicators (KPIs).

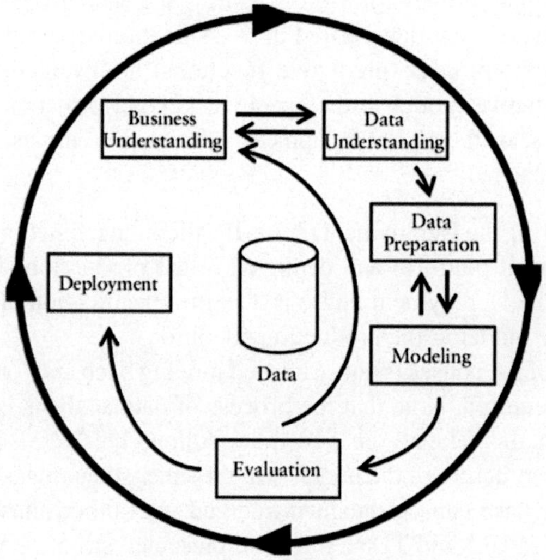

Fig. 1. CRISP-DM methodology.

Data Preparation. The data preparation phase covers all activities to construct the final dataset (data that will be fed into the modelling tool(s)) from the initial raw data. Data preparation tasks are likely to be performed multiple times, and not in any prescribed order. Tasks include table, record, and attribute selection as well as transformation and cleaning of data for modelling tools. In order to collect initial raw data from the domain environment, data scientists have to cooperate with the domain IT-specialists responsible for the setup of the data integration components.

Modelling. In this phase, various data modelling techniques (such as decision trees, neural networks, logistic or linear regression models, etc.) are selected and applied, and their parameters are calibrated to optimal values. Typically, there are several techniques for the same data analysis problem type. Some techniques have specific requirements on the form of data. Therefore, stepping back to the Data preparation phase is often needed.

The result of this phase is the set of data analytics models (i.e. predictive functions) that appears to have high quality from a data analysis perspective. This quality is usually evaluated by applying the model to the independent testing data set not used during the building of the model and computing the evaluation statistics such as classification accuracy (number of tested records with the correctly predicted value by the model/total number of records in the testing dataset) etc.

Evaluation. Before proceeding to final deployment of the model, it is essential to more thoroughly evaluate the model, and review the steps executed to construct the model, to be sure it properly achieves the business objectives. Within this phase, the quality of the models evaluated on the testing dataset is projected into the business-oriented key-performance indicators specified in the Problem understanding phase. For example, during the Problem understanding phase, domain experts estimated that if some defect in the product can be detected beforehand during the production process, it is possible to save a specified amount of waste materials. During the Modelling phase, data scientists built a predictive function and estimated its classification accuracy to 98%. During the Evaluation phase, the models (predictive functions) are evaluated by computing the impact of the predictive model to the specified key-performance indicators from the evaluation statistics. At the end of this phase, a decision on the use of the data analysis results should be reached.

Deployment. During the Deployment phase, the predictive function is packed, exported from the development platform and deployed in the production site environment. An important part of the Deployment phase is also the specification of the monitoring and maintenance of the model in the production conditions.

The sequence of the phases is not strict, and moving back and forth between different phases is always required. Note that the process of data analysis is generic and can be used in various domains. The presented division follows the CRISP-DM standard, which describes the basic structure of the process. In the other standards, some new phases can be introduced, or phase can be renamed, merged with other phase or excluded. New standards such as ASUM and TDSP integrate presented structure with the principle of the agile software development and extends the data-analytical process to the whole life cycle of the software components which integrate deployed data-analytical models.

2.2 Semantic Modelling of Data Analytical Processes

The successful implementation of the data-analytical processes requires effective communication between the domain experts and data scientists.

The most knowledge and communication-intensive phases are mainly *Problem understanding* and *Data understanding* phases where the data scientists need to understand the overall goal of the data analysis, and which data are available. In opposite, during the *Evaluation* phase, the domain experts need to communicate with the data scientists in order to understand the results of the data analysis and evaluate the impact of the data-analytical models.

This requires formalization of the business or research goals in the form of the measurable KPIs. Subsequently, the goals are transformed into the tasks of the data analysis by the data scientist who will also map domain-specific KPIs to the technical quality indicators, which can be estimated for the data-analytical models. Another part of the shared domain knowledge is the information about the available data elements and known relations between them. This can include known relations between the input data and the target value (i.e. prediction) or any dependency between multiple inputs. Similarly, for the KPIs, the data dependency can be only partial or indirect.

In the current practice, the communication between the domain experts and data scientists is documented mostly in the textual non-structured or semi-structured documentation. The documentation in natural language causes all problems related to the terminological heterogeneity and ambiguity, which leads to inefficient communication where many domain concepts have to be redefined multiple times. Another disadvantage is that the textual description of relations between the data elements or between the KPIs and technical indicators is not machine-readable, so it is not possible to automatize the common data-analytical tasks and evaluation of the results. This can also cause subsequent problems with the reproducibility of the results.

Formalization of the data analytical process in machine-readable format can also be used for the automation of the process. Typically, data scientists implement data pre-processing, modelling and validation phases as the set of scripts in the programming language such as R or Python and dedicated set of libraries. These software development tasks are very resource consuming and error-prone. On the other side, most of the operations over data are determined by the type of data and algorithm used for data processing. When these constraints are formally described, it is possible to use automatic reasoning to infer which algorithm is suitable for the specified task, how to evaluate the output model, and from the description of input data and algorithm to infer which data pre-processing operations are needed in order to prepare data for the specific algorithm. Such automation can substantially reduce the manual effort of the data scientist who can concentrate on the phase of data and problem understanding, which is the most domain-specific and knowledge critical.

2.3 Technologies for the Semantic Description of Data-Analytical Processes

Although the data analysis methodologies are general and can be applied in various application domains, they are not fully formalized. In order to support reusability and automation of the data analysis processes, the various machine-interpretable formalisms were proposed for workflow modelling. The formalisms can be divided into general formalisms for the process and workflow modelling and formalisms specific to the data analysis processes.

General standards for process modelling include the Unified Modelling Language (UML) activity diagrams or Business Process Model and Notation (BPMN) models. BPMN was initially designed as the graphical notation for modelling and exchanging of business processes where activities can be performed automatically by some software components or manually by humans. The graphical notation was later extended with the specification of the execution semantics, and the models can be directly interpreted by workflow engines which can coordinate human and machine activities. Besides the established industrial standards, many workflow engines implement their proprietary formalisms and domain-specific languages, but the main problem with the proprietary implementations is the interoperability and ambiguity in the execution semantics.

Data analysis formalisms include standards for exchanging of predictive models such as XML-based Predictive Model Markup Language (PMML) or more general Portable Format for Analytics (PFA). The data analysis formalisms are oriented towards the interoperability and deployment of the data analysis model into the operating conditions or for validation. The formalisms include constructs for the description of the input and output data (i.e. predictions), transformation functions (i.e. functions for data normalization, filtering, record and attribute selection, etc.) required for the pre-processing of the inputs/outputs and scoring/application functions of the data analysis models (i.e. neural networks, decision trees/rules, linear or logistic regression etc.) with the encoding of their parameters.

Multiple semantical models formalized in the form of the ontologies were proposed for the description of the data-analytical processes. The base model is OntoDM ontology [2], which is internally divided into three modules. The first module is dealing with the representation of the input and output data, and it is based on the ISO norm for the description of data types (atomic or compound). The second module defines concepts for the description of data sets, data-analytical tasks (i.e. classification, clustering, etc.), data-mining algorithms and the outputs of the analysis in the form of the generic models. The third module combines the previous concepts for the top-level description of the whole process according to the CRISP-DM methodology.

An essential part of the OntoDM ontology is the specification of the constraints, which must be fulfilled by the final data-analytical models. The constraints are specified for the input data (i.e. types of the attributes, missing or extreme values, etc.), the complexity/interpretability of the models (e.g. produced models have to be in the form of rules with the limited number of condition terms etc.) and for the performance of the models (e.g. minimum precision estimated on the testing dataset or using the cross-validation).

Other proposed ontologies further extend the definition of some OntoDM concepts. The DMOP ontology [3] is focused on the detailed description of the data mining algorithms, including the description of their internal structure and principles, such as the description of the numerical optimization or regularization methods. The DMWF ontology [4] defines dataflow operators, which can be applied for data pre-processing, modelling, and evaluation. Each operator is described by the set of required inputs and outputs and by the pre- and post-conditions. The pre- and post-conditions are specified as logical expressions in SWRL language, which allows using of automatic reasoning for the composition of the data-analytical workflows.

Another generic proposal which extends OntoDM ontology is Expose [5]. The Expose ontology extends the phases of CRISP-DM process for the generic description of scientific experiments in order to achieve reproducibility and reusability of the results. Besides the conceptualization, Expose also provides the XML based language for the publication and exchange of the experiment descriptions in a machine-readable format.

The conceptualization of the described semantic models is based on the description of programming interfaces of existing software tools such as R or Python scikit-learn environment. Another relevant technology is the standardized formats for the exchange of data-analytical models, such as PMML XML based standard, PFA format based on JSON notation or ONNX standard focused mainly on the exchange of deep learning models.

3 Semantic Framework for Modelling of Data-Analytical Processes

In [6] we have proposed the unified semantic framework for the modelling of the data-analytical processes which covers all phases. The framework has a modular architecture and can be divided into the following modules:

- *Domain concepts* – this module specifies concept which describes entities and their relations from the domain in which the data analysis methods are applied.
- *Key performance indicators and data elements* – this module contain the concepts for the specification of KPIs which formalize the business and research goals of the data analysis and the concepts for the description of the input and output data elements and data sets.
- *Algorithms and data-analytical models* – this module specifies concepts for the description of data mining algorithms, their settings and concepts for the description of the analysis results in the form of the data mining models.
- *The process model for data-analytical workflows* – this module contains concepts for the description of data pre-processing operators and the composition of workflows.

The first module is designed from the perspective of the domain expert, which formalizes the analyzed domain using the domain concepts. The second module contains concepts which are shared between the domain expert and data scientist. The last two modules are primarily used for the automatic composition of the data-analytical workflows and the automation of the overall data-analytical process.

3.1 Doman Concepts

Concepts for the formalization of the domain-specific knowledge are specified using the SKOS meta-model, which allows defining unique label and narrative description of the concepts localized in multiple natural languages. Domain concepts can be arbitrary hierarchically organized in the form of thesaurus/taxonomy using the narrower/broader relations. It is also possible to define the poly-hierarchical schema, where one concept can have broader parents. Besides the hierarchical relations, the concept can also be linked using association relations. A controlled vocabulary of the domain concepts can also be used as the classification schema for the organization of the various types of the documents which can be part of the domain documentation, or which can be created by the data scientists in order to document the data analytical process. SKOS taxonomy can also be used to describe types of documents.

Besides the narrative description and conceptualization, in some domain, it is suitable to describe and explain existing concepts in the form of diagrams, schema or other types of the graphical notations (e.g. using the BPMN diagrams for the modelling of business processes). In this case, graphical elements are described semantically as the SKOS concepts, which are linked with the corresponding concept in the domain model. This allows implementing two-way navigation between the formalized set of semantic concepts and the graphical elements in the diagram. However, the proposed framework does not specify how these links are represented in the semantic representation or embedded in the file format of the graphical notation.

3.2 Key Performance Indicators and Data Elements

Data elements are modelled in two levels of abstraction: logical and physical (see Fig. 2). Logical *DataElement* concept describes the role, type and quantity (e.g. units of measurements) of any data element related to the production step or equipment. Data element can have an input role (for measurements or input control signals) or output role (for diagnostic signals). Output data elements can be selected as a target attribute for the predictive function. One data element can have both roles depending on the goal of the data analysis (i.e. control signal can be an input for the predictive maintenance or output for the predictive control). The type of the data elements characterizes elements according to their values and denotes continuous, ordinal, nominal, spatial or series data elements.

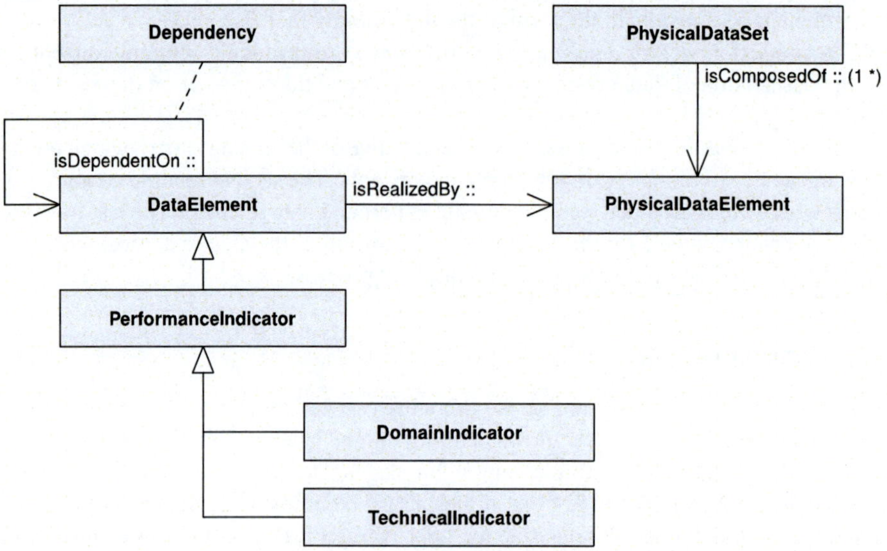

Fig. 2. Concepts representing data elements and Key Performance Indicators.

Physical Data Elements (concept *PhysicalDataElement* in Fig. 2) link logical Data Elements to the physical representation of the data in the structured records or files. Multiple Physical Data Elements (fields) can be grouped into one record or file and described by *PhysicalDataSet*. Each Physical Data Element has a specified data type. The data type can be primitive (e.g. byte, int, string) or complex (array, enumeration, map or union of types).

The concepts of Key-Performance Indicators (KPIs – *PerformanceIndicator* concept in Fig. 2) represent metrics designed to visualize, assess and manage the performance or impact of specific operations within the process industries. The main properties of the KPI metrics (e.g. type and quantity) are inherited from the Logical Data Elements. The KPIs can be grouped into KPI categories according to the classification in ISO 2240.

Besides this classification, causal relations between KPIs can be expressed by the KPI Dependency relations, which transform one KPI to another (e.g. material or energy savings can be converted into KPIs for the environmental impact such as level of emissions). The KPI concepts are divided into the *DomainIndicator,* which describes target domain-specific KPI and *TechnicalIndicator*, which describes the performance of the data mining models and the complexity of the models. The values of Technical Indicators are usually estimated using the testing the model on the independent test dataset. Using the composition of KPI Dependency relations, it is possible to infer the impact of the deployment of a predictive function on the performance of the production process, where the performance can be evaluated from various perspectives such as material/energy consumption, product quality, environmental impact, etc.

3.3 Algorithms and Data-Analytical Models

The basic concepts for the representation of the data mining algorithms and data analytical models are depicted in Fig. 3. The main type for the representation of algorithms is *Algorithm* class, which extends the *Operator* class. The *Operator* class represents general operation, which transforms inputs of the given type to the requested output (i.e.

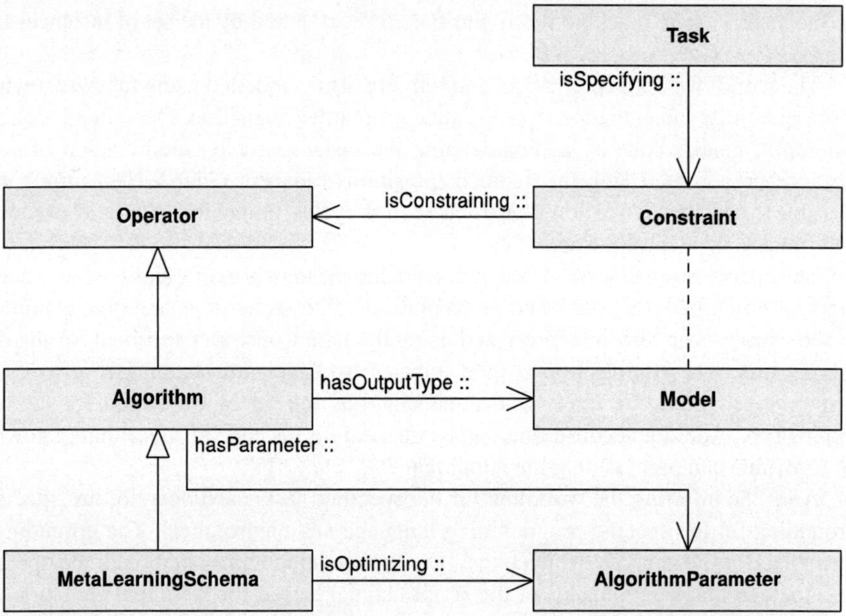

Fig. 3. Concepts representing data mining algorithms and data-analytical models.

in the case of the data mining algorithms, operator, which learn a data-analytical model from the training dataset). The outputs of the algorithms are generalized data-analytical models represented by the *Model* concept. Some types of the models can be directly applied to the data (e.g. classification or regressions models), which means that the specific types of the data mining models can also be directly extended from the *Operator* concept.

Besides the types of output, data mining algorithms are associated with the set of (hyper)parameters, which must be specified by the data scientist. The data mining task is formally defined by *Task* concept, which specifies the set of constraints (class *Constraint*) and defines requested properties of the generated data-analytical model.

The constraints are specified using logical expressions. Constraints cover characteristics of the input data (e.g. types of the data elements, the occurrence of missing/extreme values, etc.); type of the model (e.g. classification, regression, clustering, association rules, anomaly detection, etc.); and quality or interpretability of the models by constraining the technical indicators.

The constraints are further divided into the so-called hard constrains, which must be strictly fulfilled; and soft constraints which should be included as the criterion for the solution but can be relaxed and the output does not strictly fulfil all soft constraints.

3.4 The Process Model for Data Analytical Workflows

The proposed process model is suited for the annotation of existing data analytical scripts and the automatic generation of workflows for the specified task. The goal of the annotation of an existing script is to achieve reproducibility and reusability of the existing workflow by formalized semantic representation. The proposed model is based on the abstract state machine where the state is represented by the set of instances (see Fig. 4).

The workflow (concept *ProcessModel* in Fig. 4) is modelled using the *Nodes* which represent either the operation over the state or guarded transition. Operations (concept *Operator*) modify state by adding/deleting the instances or by modification of some instance properties. Using the guarded transitions (concept *GuardedTransition*), it is possible to specify control flow constructs such as cycles, branching or parallel execution and synchronization of the activities.

The operators are described functionally using the logical expressions, which specify inputs, output, pre-conditions and post-conditions. Pre-conditions and post-conditions of the operator can also be represented using the type Constraint specified for the data mining task. The composition is then inferred by backward chaining of effects and assumptions, where the target goal is specified as the set of Constraint for the data mining task. Additional constraints can be inferred for the specific algorithm (e.g. when an algorithm can process only the numerical data, etc.).

After the inferring the workflow for the specified task, each operator has specified grounding for the specific programming language and environment. The grounding is formalized as the template which transforms the semantic representation of the operator into the executable commands of the script language used for the implementation and execution of the final generated workflow. The operators can have assigned multiple grounding, which allows to generate semantically equivalent workflow into the multiple

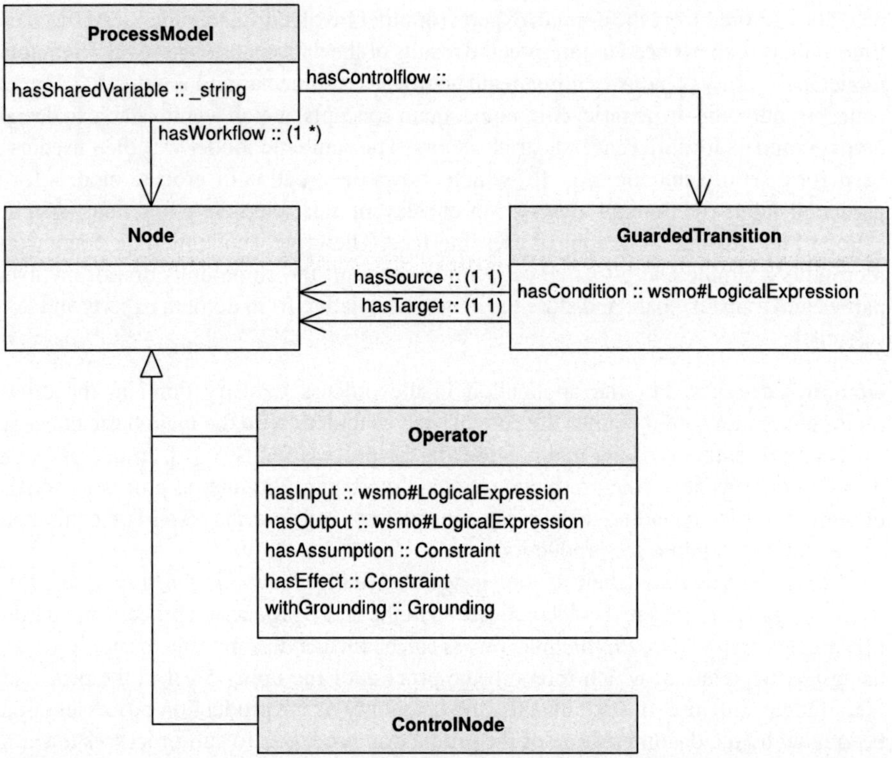

Fig. 4. Concepts representing the overall process of the data analysis and data analytical workflows.

target programming environments such as R or Python. In this way, the semantic model also supports the migration of the data analytical scripts from one environment to another one.

4 Semantic Data Analysis Case Studies

In the following sub-sections, we describe the implementations of the proposed semantic framework for modelling of data analytical processes in two different domains - process industries and network intrusion.

4.1 Process Industries Use Case

The primary motivation to design and develop of the cross-sectorial semantic model in process industry domain was to provide a common communication language between the domain experts and data scientists when solving data analytical tasks in particular industry area [7]. When solving a data analytical task, data scientists require extensive knowledge of the specifics related to the production processes and underlying data, as well as a good understanding of the business objectives of the task. Such knowledge could be acquired from the domain experts (or other involved stakeholders). At the same time, domain experts need to interpret the results of the data analytical models. Semantic model in the process industry domain did not focus on some particular industrial domain concepts, but rather to generic, common domain concepts, which would enable to design domain models for different industrial sectors. The semantic model was then used as a basis for a set of semantic tools [6] which allows the creation of process models for a particular industrial domain, description of relevant data, predictive functions, specification of KPIs using the graphical user interface. Therefore, semantic tools framework represents a simple and effective way, how to support the communication of involved parties and transfer, share and store the expert knowledge from domain experts and data scientists.

Domain Concepts. For the application in the process industry domain, the cross-sectorial domain model defines the core of the vocabulary with the main basic concepts for the representation of the main entities in the process industry [6]. In this case, the domain concepts cover the production process modeling. Production processes consist of segments (corresponding to the production phases) and describe resources, equipment and personnel required for production.

Domain concepts and their relation is depicted in Fig. 5. *ProductionProcess* and *ProductionSegment* represent the decomposition of the entire production process into particular process steps. Also, *ProductionProcess* can be further decomposed to sub-processes using *partOf* relation, which refers to the process of the upper level in the processes hierarchical structure. In such manner, the taxonomy of the production processes could be created by the decomposition of the production processes to sub-processes (each of the sub-processes could be modelled using process segments).

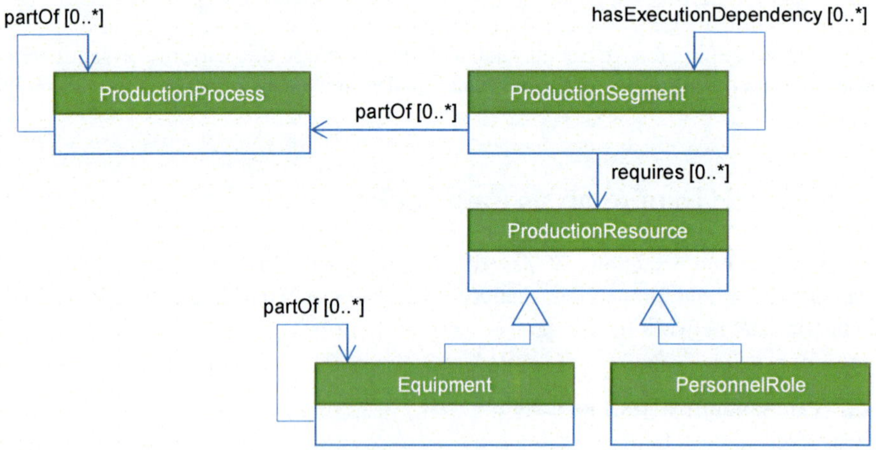

Fig. 5. Concepts representing the overall process of the data analysis and data analytical workflows.

The *ProductionSegment* is a logical grouping of resources (represented by a *ProductionResource* concept) required to perform a particular production process step. Production resources are then divided into *Equipment* and *PersonnelRole* concepts specifying what personnel is necessary to carry out a specific activity within the process and what type of equipment is being used. *Equipment* covers any equipment-related entities, e.g., production sites and areas, production units, production lines, work, and process cells and may enable to include other equipment (e.g., production line consisting of machines and devices, each of them modelled separately). *PersonnelRole* describes any human roles involved in the production process and describes their capabilities. Both equipment and personnel concepts can be organized in concept hierarchies. Additionally, each *ProductionSegment* can be annotated with the relevant data element concepts describing its parameters and capabilities, or describing the data related to the resource (e.g., sensor data obtained from the production machine).

Production segments are ordered into the execution workflow by *hasExecutionDependency* relationship. It is used to specify a condition when certain *ProductionSegment* can be executed only after another *ProductionSegment* has been finished.

Semantic Modelling in Aluminium Production Processes. The semantic model was used in the domain of aluminium production. Aluminium production is performed using electrolysis in the potline. The potline consists of a set of several hundred pots where liquid aluminium is being produced using electrolysis. The main inputs to the process are electricity, alumina, anodes, and an electrolytic bath; each pot is equipped with anodes and cathodes. Anodes and their quality are some of the most critical and controllable inputs for the electrolysis.

The proposed semantic model enabled to create and define the aluminium production process model and to specify the process taxonomy and particular production segments within the processes.

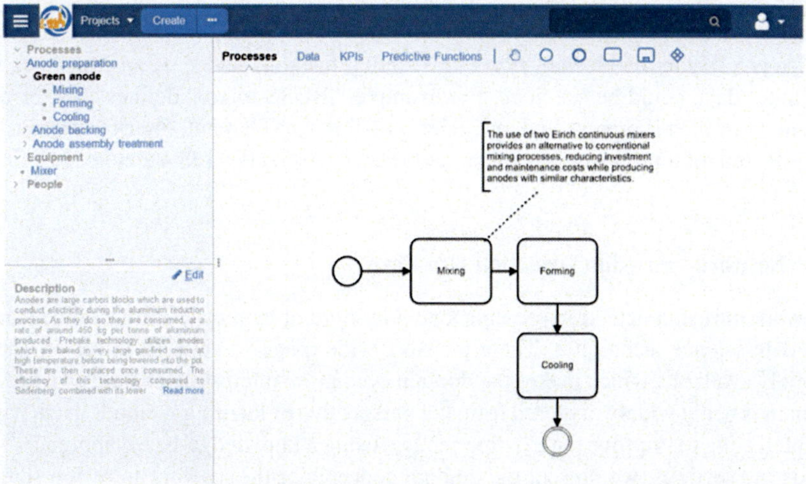

Fig. 6. Simplified process model of anode production.

When using Semantic Modeler [6] as a tool built on the semantic model, simple GUI was used to define the process segments and build the overall process workflow. The example of a simple process model (anode production process) as depicted in Fig. 6. Such easy-to-use interface enabled the domain experts with no prior knowledge of data science to input the essential concepts - processes, divide them into the process segments, annotations, etc. and therefore create a common knowledge base during the problem and data understanding phase of the data analytical process.

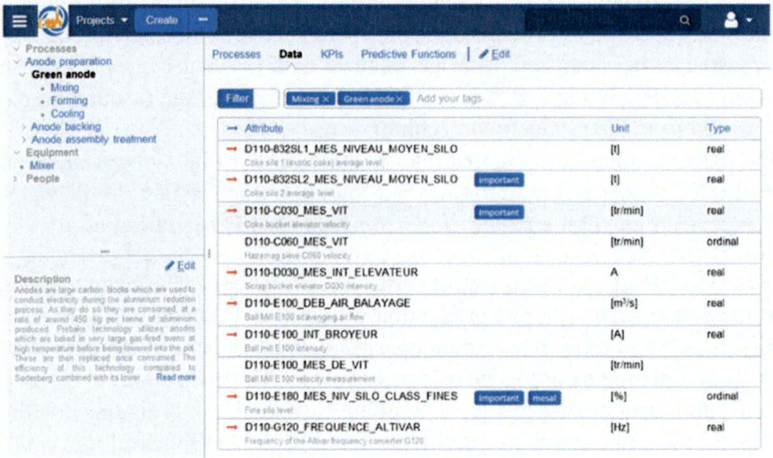

Fig. 7. Data elements of the anode production.

In a similar fashion, relations with other concepts can be specified. Domain experts can specify the particular devices involved and their relation to the process segment (e.g., define the "BUSS mixer" equipment and its relation to the "Mixing" production segment). Also, a concept hierarchy for the devices can be created (e.g., define that "Mixer" device is a "Hydraulic device" type). In a similar fashion, device parameters, e.g., input or output data, could be specified. For example, "BUSS mixer" defines a mixer power parameter, which is represented by a corresponding data element. Figure 7 demonstrates the selection of a production segment and related parameters (data elements).

4.2 Network Intrusion Detection Use Case

Network intrusion detection presents a popular field of application of machine learning and data science techniques. There is also a wide range of network intrusion datasets publicly available, which makes the domain even more interesting and attractive. As the domain is being widely explored from the perspective of learning methods used, it could be interesting to improve the existing models using a knowledge-based approach. In the following sections, we present the domain concepts of the network intrusion semantic model. The model captures the essential concepts related to security and network attacks. The model is then used to select the most suitable model with respect to a type of prediction, we would like to achieve (e.g., if we want just to detect the attacks, or to predict the concrete attack type), or to extract the domain-specific knowledge, not explicitly contained in the data, which could enhance the predictions.

Domain Concepts. The domain concepts in the proposed network intrusion semantic model consist of the most important terms in the network intrusion domain. The core structure of the domain concepts is depicted in Fig. 8.

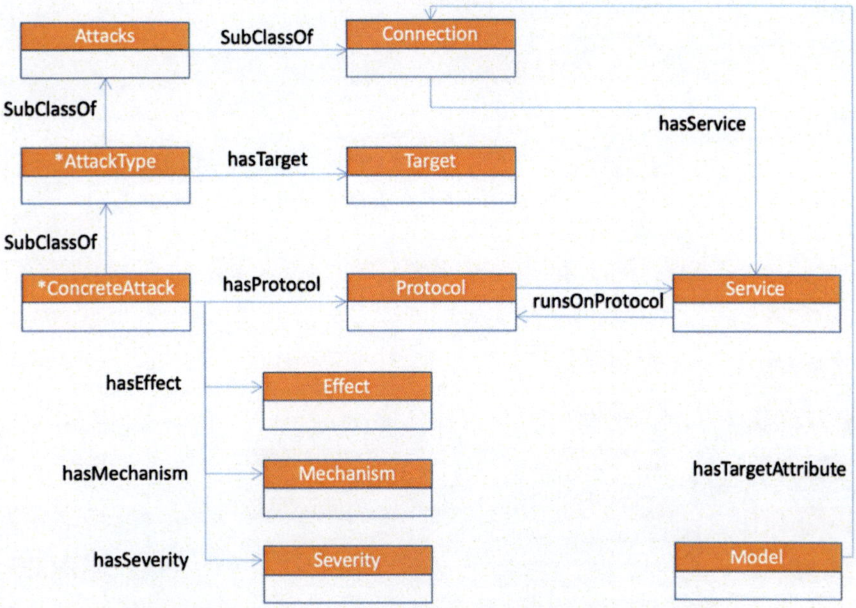

Fig. 8. Main concepts of the network intrusion domain model.

The network intrusion domain model consists of these concepts:

- *Connection* represents each connection and consists of multiple sub-classes that define a concept hierarchy. Connections are further divided into Attacks representing network attacks and the Normal, non-attack records
- *Effect* defines all possible consequences of individual attacks (e.g., slows down server response, execute commands as root, etc.)
- *Mechanisms* represent all possible causes which could lead to particular attacks (e.g., poor environment sanitation, misconfiguration, etc.)
- *Protocol* represent the types of protocols on which the connection (attack or non-attack) is running (e.g., TCP, UDP, or ICMP)
- *Service* represent each type of connection (e.g., http, telnet, etc.…)
- *Severity* represents the severity level of the attack, (e.g., weak, medium, and high).
- *Target* represent possible targets of a given type of attack (e.g., user, network). Each type of attack is aimed at a specific type of targets.

- *Model* represent individual trained and serialized predictive models. Models are represented by their URI, and each model instance addresses prediction on a specified level of Connection concept taxonomy.

The crucial part of the knowledge model is the Attacks concept taxonomy. The taxonomy defines the types of attack and concrete attacks of the specified kind. The taxonomy was created using data from an available dataset from the network intrusion domain and is depicted in Fig. 9.

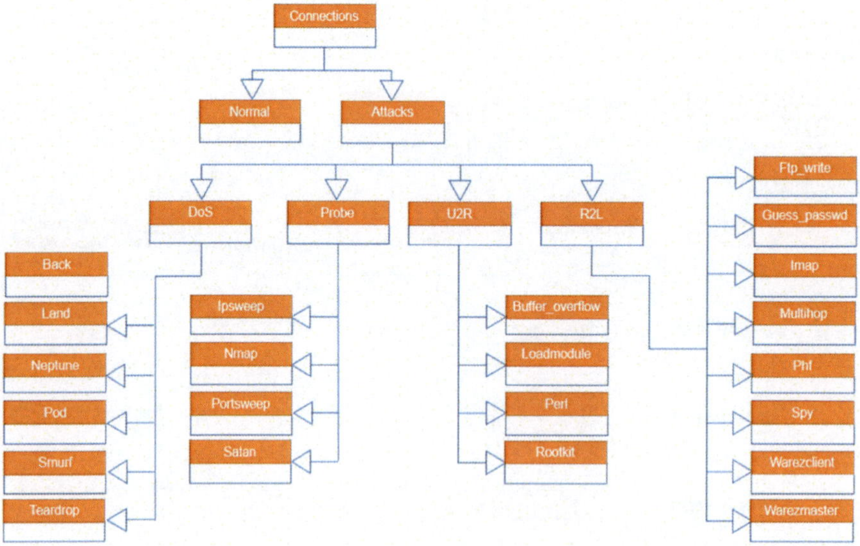

Fig. 9. Main concepts of the network intrusion domain model.

Network Attack Prediction Using the Semantic Model. The semantic model was used in hierarchical knowledge-based network intrusion detection system (IDS) presented in [8]. The ontology was implemented in OWL language; predictive models were trained using Python data science tools (scikit-learn). As the standard data science Python stack was used (Pandas, Numpy, scikit-learn), to communicate with the ontology, we used the RDFlib package, which supports querying of the ontology using SPARQL queries.

Different predictive models (and using multiple machine learning algorithms, e.g., Random Forests) were trained to detect the network attacks on different levels of target attribute generalization (corresponding to *Attacks* taxonomy from the ontology) and were serialized and instantiated in the ontology as instances of the *Model* class.

When classifying an unknown connection, hierarchical knowledge-based IDS in the first step checks, whether a connection represents an attack or a possible threat. Therefore, IDS need to employ a predictive model capable of separation of *Connection* records into *Normal* and *Attacks* classes. IDS use a SPARQL query to retrieve the suitable predictive model and runs a prediction. If a prediction signalizes an attack record, IDS follows with the prediction of the attack type. In a similar fashion, a SPARQL query retrieves the predictive model capable of prediction of concepts on the following level of *Attacks* concept hierarchy. The same steps are repeated to predict the concrete attack of a given type.

To demonstrate the contextual, domain-specific information, IDS is able to provide the domain-specific information which could extend the predictions, e.g., in case, when the system is not able to predict the concrete attack. In such a scenario, the model could reliably predict the attack type. To better specify the prediction, the system could retrieve the predicted severity level of the attack type, which may help to determine the possible consequences of the attack.

5 Intelligent Data Analysis in the Medical Domain

We have shown in the previous section that for technical applications like process industry or network intrusion detection there is a quite significant portion of background knowledge available in explicit form and we presented how to make use of it by means of our semantic framework for modelling of data analytical processes. However, the situation is often quite different in some other domains like medicine, where a significant portion of background knowledge is in tacit form.

5.1 Knowledge Extraction and Use in Medical Analysis

In the current practice, the interaction between the domain experts and data scientists is realized through various communications channels like textual non-structured form (emails, short messages), semi-structured documentation (reports, literature) or verbally. Each of this channel has its advantages and disadvantages. Sometimes, it lacks the common vocabulary to prevent possible misunderstandings and inefficient directions of the experiments.

The collaboration between domain experts and data scientists in healthcare area requires a lot of communication and interactions to ensure correct interpretation of the input data, the goals definition meeting the expectations and evaluation of the obtained results from the all relevant aspects. A summary of the main tasks of this collaboration is depicted in Fig. 10.

In many cases, the results are represented by a set of various graphs, decision models or tables. It is necessary to effectively narrow the search space and provide the results in a simple, understandable and intuitive form. For this purpose, we used several combinations of methods from statistical analysis, machine learning, data mining or exploratory data analytics. Selected case studies are shortly presented below. But before showing particular cases, let us have a more detailed look at how the background knowledge can be captured in medical data analytical processes.

An explicit part of the knowledge can be represented similarly as previously, by means of suitable semantic representation (domain concepts). For efficient support of tacit knowledge available in brains of the medical experts, a suitable form of interactive decision support system is needed.

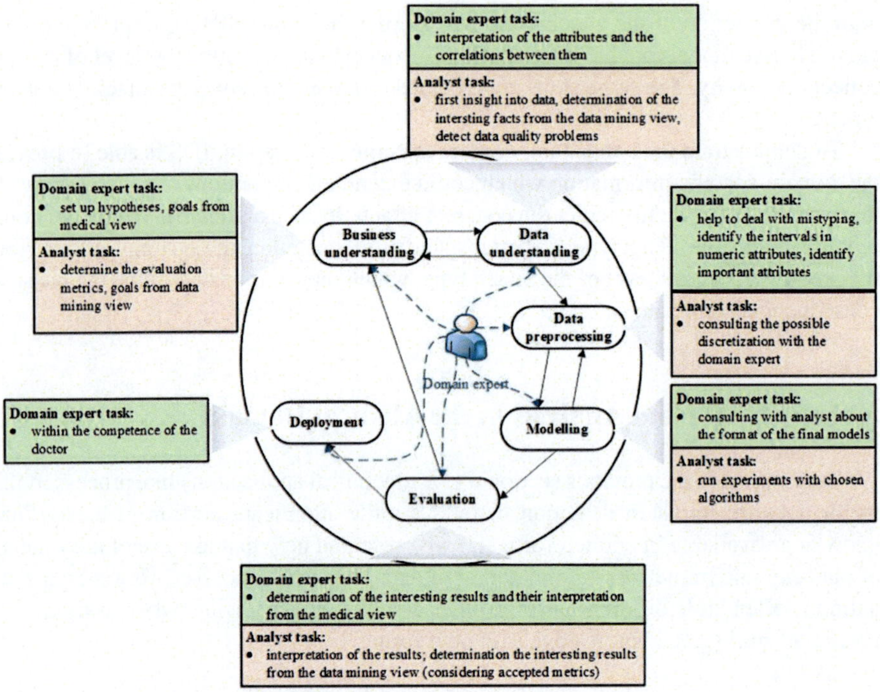

Fig. 10. CRISP-DM methodology with the human-in-the-loop concept [20].

Domain Concepts. The electronic health care records (EMR) represent a new opportunity on how to improve the quality of the medical diagnostics and the overall healthcare. But this opportunity also brings some barriers for the routine daily use of this data like a limited scope of the collected data, lack of compatibility between different systems, non-uniform terminology and missing connections with other sectors or databases. The semantic representation of the EMRs is important mainly from the point of view of digital interoperability between the various healthcare information systems. Additionally, formalized medical data allow better interpretability of the clinical cases and better reproducibility of the medical trials.

We have developed a system for the extraction of information from electronic health records of cardiology patients [24]. There are altogether extracted 64 attributes signed by the expert as relevant for the assessment of cardiovascular risk. Currently, a larger project is running [25], where a much broader number of relevant information will be collected and researched in order to contribute to early cardiovascular risk detection by means of data analytics processes.

From the point of view of semantic formalization, the medical domain has well established standard ontologies which include Unified Medical Language System (UMLS), International Classification of Diseases (ICD) or Systematized Nomenclature of Medicine Clinical Terms (SNOMED CT). Terminological ontology SNOMED-CT is particularly suitable for the description of the data analytical tasks and the formalization of the communication between the domain experts (clinicians) and data scientists.

SNOMED-CT is a concept-oriented and machine-readable medical terminology used in electronic medical records. The clinical descriptions in practice which are usually documented as the semi-structured documents, can be transformed into the SNOMED CT concepts and descriptions. Besides the existing concepts, the SNOMED CT ontology can be easily extended by clinicians with the new formalized terminology in order to cover in the knowledge base whole data-analytical process.

Decision Support System. In order to mediate the interaction of domain experts possessing valuable tacit knowledge with a system based on data analytical process, the human-in-the-loop principle [21] is very valuable. The main idea behind this concept is that data mining algorithms can interact with agents, and based on this interaction, optimize its behaviour. This is also known as interactive machine learning.

One of the possible approaches on how to implement this idea into the data analytical system is to use case-based reasoning (CBR) principles. CBR is an established problem-solving strategy which tries to adopt the solution of previous similar cases to the current one. This is a natural and often used strategy in medical science. One particular case study where we used these principles will be presented in subsection

5.2 Metabolic Syndrome Diagnostics

In [9–11] we used a combination of clustering and logistic regression to extract patients' groups with similar characteristics, medical history or symptoms; with the following step to identify the main differences between these groups to support the decision process of the primary care doctor or specialist.

The typical analytical process starts with a common understanding between data analysts and domain experts what are the expectations that can be transformed into goals and success criteria. This phase follows by basic data understanding using suitable statistical tests or data exploratory methods like histograms, correlation analysis, etc. We extended this phase with an estimation of new, possible useful cut-off values for selected variables through the Youden index method [12]. In comparison to the Receiver operating characteristic curves, this method offers better results with respect to the maximum overall correctly classified cases by maximizing the sum of sensitivity and specificity metrics. It helped us to identify more optimal points above which the greatest number of women with Metabolic Syndrome can be expected, with the smallest number of false negative results. The International Diabetes Federation provided in 2005 the definition of Metabolic Syndrome for the females such as a set of rules containing waist circumference, hypertension, triglycerides, cholesterol, and fasting glucose. We used the decision tree approach to confirm this definition in the dataset in parallel to extract other simple, understandable diagnostic rules.

The prevalence of multimorbidity or multiple diseases in the same person increases with age and is partly due to the synergistic effects that different diseases have on each other [13]. In [10], we aimed to identify the clustering of comorbidities, cognitive, and mental factors associated with increased risk of pre-frailty and frailty in patients older than 60 years. For this purpose, we used the k-Means algorithm [14]. It is a type of partitioning algorithm that uses only numerical parameters and aims to partition a set of n objects into k clusters so that the resulting intra-cluster similarity is high, and the

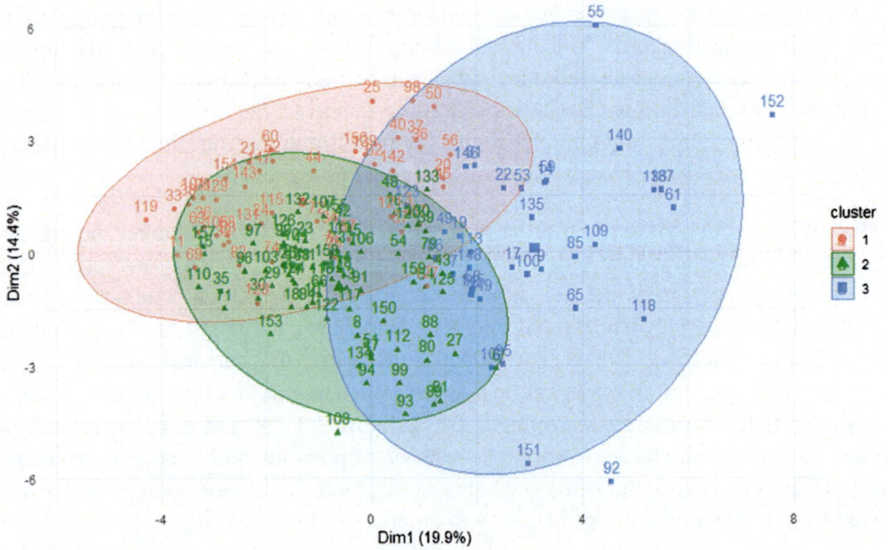

Fig. 11. Scatter plots of the three generated clusters in the Metabolic Syndrome case study.

inter-cluster similarity is low. The intra-cluster similarity is measured with the mean value of distances between the objects in the cluster. As the initial value of k is very important, we used the Elbow [15] and Average silhouette methods [16] but without unambiguous results. Finally, we applied the Calinski-Harabasz index [17] and decided to generate 3 clusters. The best R2 value indicating the low inter-cluster distance was 0.19, see Fig. 11.

The most important factors for the frailty diagnosis were identified by the logistic regression models [18]. At first, we checked all input variables by simple correlation analysis and variance inflation factor to prevent the multicollinearity [19].

5.3 Cardiovascular Risk Assessment

Cardiovascular diseases hold for long a leading position in statistics on the causes of death in many countries. The global trend in the proportion of deaths from these diseases is constantly growing. One way how to make procedures for early and precise assessment of cardiovascular risk more effective is to use a decision support system to assist cardiologists in making decisions about the severity of a patient's condition.

We have designed and developed a decision support system based on CBR principles and making use of data analytical process at the same time. CBR is a methodology used to solve problems, based on previous cases with a known solution. This method is based on the principles of human reasoning, trying to imitate a person in solving new problems, who uses his previous experience and tries to adapt them to the current situation [22]. Solving a problem with CBR usually consists of four main phases [23]:

1. RETRIEVE - finding the most similar cases to a new problem,

2. REUSE/ADAPT - reuse of knowledge from found cases to solve a new problem,
3. REVISION - review of the proposed solution,
4. RETAIN - saving parts of the solution that may be useful in the future.

In our approach we adopted this CBR methodology for the assessment of the cardiovascular risk of new patients, utilizing the k-NN method for finding the most similar patients from the history (within the RETRIEVE phase) and an original method for reusing the information about the similar cases in order to assess the cardiovascular risk of the new patient (within the REVISE phase). Moreover, the CBR cycle is supported by a set of visualizations, which have been designed in tight cooperation with the target user, a cardiologist. Based on his feedback on the first version of our system, it seems to be very useful and promising. More details can be found in [26].

6 Conclusions

In this chapter, we have described the semantic model for the formalization of the data analytical processes. This ontology promotes the sharing of best practices and results within and between domain experts and data scientists, reducing both the duplication and loss of knowledge. It is an essential step in formalizing the data analytics methods and how they can be applied in the various domains. The described semantic model has modular architecture divided into a domain-specific part and data analytics part. The data analytics part specifies concepts for data representation, pre-processing, and modeling of the predictive functions. The domain-specific part specifies the conceptualization of the modelled domain. In the chapter, we have presented two use cases: one for the optimization in the process industries and one for the intrusion detection systems. Both parts are interlinked in one unified semantic framework. Besides the formal specification of the semantic models, we also designed and implemented modelling tools that allow collaborative semantic modelling and sharing of knowledge between domain experts and data scientists. An essential feature of the modelling tools is that they are designed for users without skills in semantic technology. In the future, we like to adopt the proposed model to other domains and provide a common metamodel which will also allow the cross-sectorial data analytical processes.

Acknowledgements. This work was supported by the Slovak Research and Development Agency under grants no. APVV-16-0213 and APVV-17-0550.

References

1. Shearer, C.: The CRISP-DM model: the new blueprint for data mining. J. Data Warehous. **5**(4), 13–22 (2000)
2. Panov, P., Dzeroski, S., Soldatova, L.N.: OntoDM: an ontology of data mining. In: 2008 IEEE International Conference on Data Mining Workshops, pp. 752–760 (2008)
3. Hilario, M., Nguyen, P., Do, H., Woznica, A., Kalousis, A.: Ontology-based meta-mining of knowledge discovery workflows. In: Meta-Learning in Computational Intelligence (2011)

4. Kietz, J., Serban, F., Bernstein, A., Fischer, S.: Towards cooperative planning of data mining workflows. In: Proceedings of the Third Generation Data Mining Workshop at the 2009 European Conference on Machine Learning (ECML 2009) (2009)
5. Vanschoren, J., Soldatova, L.: Exposé: an ontology for data mining experiments. In: International Workshop on Third Generation Data Mining: Towards Service-Oriented Knowledge Discovery (SoKD-2010), pp. 31–46 (2010)
6. Sarnovsky, M., Bednar, P., Smatana, M.: Cross-sectorial semantic model for support of data analytics in process industries. Processes **7**(5), 51–68 (2019)
7. Sarnovsky, M., Bednar, P., Smatana, M.: Big data processing and analytics platform architecture for process industry factories. Big Data and Cognitive Comput. **2**(1), 3 (2018)
8. Sarnovský, M., Paralič, J.: Hierarchical intrusion detection using machine learning and knowledge model. Symmetry, **12**(2) (2020)
9. Sabanovic, S., Majnaric Trtica, L., Babič, F., Vadovský, M., Paralič, J., Vcev, A., Holzinger, A.: Metabolic syndrome in hypertensive women in the age of menopause: a case study on data from general practice electronic health records. BMC Med. Inf. Decision Making **18**(1), 1–24 (2018)
10. Bekic, S., Babič, F., Filipčic, I., Majnaric Trtica, L.: Clustering of mental and physical comorbidity and the risk of frailty in patients aged 60 years or more in primary care. Med. Sci. Monitor **25**, 6820–6835 (2019)
11. Babič, F., Majnaric Trtica, L., Bekic, S., Holzinger, A.: Machine learning for family doctors: a case of cluster analysis for studying aging associated comorbidities and frailty. In: Holzinger, A., Kieseberg, P., Tjoa, A., Weippl, E. (eds.) Machine Learning and Knowledge Extraction. CD-MAKE 2019. Lecture Notes in Computer Science, vol 11713. Springer, Cham (2019)
12. Yin, J., Tian, L.: Optimal linear combinations of multiple diagnostic biomarkers based on Youden index. Stat. Med. **33**(8), 1426–1440 (2013)
13. Barnett, K., Mercer, S.W., Norbury, M., et al.: Epidemiology of multimorbidity and implications for health care, research and medical education: a cross-sectional study. Lancet **38**, 37–43 (2012)
14. Hothor, T., Everitt, B.S.: A Handbook of Statistical Analyses Using R, 2nd edn. Chapman and Hall/CRC, Boca Raton (2009)
15. Kodinariya, T.M., Makwana, P.R.: Review on determining number of Cluster in K-means clustering. Int. J. Adv. Res. Comput. Sci. Manage. Stud. **1**(6), 90–95 (2013)
16. Rousseeuw, P.J.: Silhouettes: A graphical aid to the interpretation and validation of cluster analysis. Comput. Appl. Math. **20**, 53–65 (1987)
17. Calinski, T., Harabasz, J.: A dendrite method for cluster analysis. Commun. Stat. Theory Methods **3**(1), 1–27 (1974)
18. Tolles, J., Meurer, W.J.: Logistic regression relating patient characteristics to outcomes. JAMA **316**(5), 533–534 (2016)
19. Habshah, M., Kumar Sakar, S., Rana, S.: Collinearity diagnostics of binary logistic regression model. J. Interdisciplinary Math. **13**(3), 253–267 (2010)
20. Lukáčová, A.: Approaches to extraction of decision support rules in medical domain. Dissertation thesis. Technical University of Košice, 99 p. (2016)
21. Holzinger, A.: Interactive machine learning for health informatics: when do we need the human-in-the-loop? Brain Inform. **3**(2), 119–131 (2016). https://doi.org/10.1007/s40708-016-0042-6
22. Begum, S. et al.: Case-based reasoning systems in the health sciences: a survey of recent trends and developments. In: IEEE Transactions On Systems, Man, And Cybernetics – Part C: Applications and Reviews, vol. 41, no. 4, pp. 421–434 (2011)
23. Choudhury, N., Begum, S.: A survey on case-based reasoning in medicine. Int. J. Adv. Comput. Sci. Appl. **7**(8), 136–144 (2016)

24. Pella, Z., Milkovič, P., Paralič, J.: Application for text processing of cardiology medical records. In: Proceedings of the IEEE World Symposium on Digital Intelligence for Systems and Machines (DISA 2018), pp. 169–174, IEEE (2020)
25. Pella, D. et al.: Possible role of machine learning in the detection of increased cardiovascular risk patients – KSC MR Study (design). Archives of Medical Science (accepted)
26. Tocimáková, Z., Pusztová, L., Paralič, J., Pella, D.: Case-based reasoning for support of the diagnostics of cardiovascular diseases. In: Studies in Health Technology and Informatics, vol. 270, NLM (Medline), pp. 537–541 (2020)

Intelligent Analysis of Data Streams

Viera Rozinajová[1], Anna Bou Ezzeddine[1], Gabriela Grmanová[1], Petra Vrablecová[1](✉), and Miriama Pomffyová[2]

[1] Kempelen Institute of Intelligent Technologies, Mlynské Nivy 5, 811 09 Bratislava, Slovakia
energo@kinit.sk

[2] Faculty of Informatics and Information Technologies, Slovak University of Technology in Bratislava, Ilkovičova 2, 842 16 Bratislava IV, Slovakia
miriama.pomffyova@stuba.sk

Abstract. The huge amount of data generated on a daily basis in many areas of our lives predetermines data science to be one of the most significant current IT research areas. Thanks to the recent technological trends, such as internet of things (IoT), the data streams constitute a majority of currently created data. This chapter aims to provide one possible view on the recent trends in stream mining.

Our focus is on two frequent data mining tasks – namely the prediction and the optimization. We have several years of experience in predictive modeling and we would like to offer here a summarization of the selected outcomes of our research work. The domain in which we have verified our methods, is the power engineering area. Due to population growth and technological advancement, there has been a huge increase of global energy demand in recent years.

The ultimate goal is to manage the energy supply in the most efficient way. To propose smart solutions, the first step is to have a clear idea about its future consumption and production. Then we can proceed to efficient control of the smart grid, by involving recent trends in optimization and utilizing machine learning approaches. Hence the second part of the chapter is devoted to our endeavors in solving tasks of smart grid optimization. The common denominator of the described approaches is the effort to cover various types of knowledge entering these procedures.

Keywords: Data analysis · Big data · Stream mining · Time series · Prediction · Optimization

1 Introduction

As we all know, the 21-st century can be characterized as a century of digital transformation. Paradigms such as *artificial intelligence (AI)*, *internet of things (IoT)*, *5G*, *high-performance computing* and *data analytics* play a crucial role – they represent emerging technologies, which will significantly change our lives. If we succeed to build the technological progress on these phenomena, respecting also ethical foundations, we will witness the formation of veritable *digital intelligence society*. It is based on several pillars – however, we are convinced, that the ability to acquire and utilize knowledge will

J. Paralič et al. (Eds.): DISA 2020, AISC 1281, pp. 98–116, 2021.
https://doi.org/10.1007/978-3-030-63872-6_5

be the key distinguishing factor of competitiveness. This fact also implies the transition of current economy to the knowledge-based economy.

We can consider the knowledge from several angles. From societal point of view its engagement in decision-making process is a significant determinant of economic growth. Our interest is much narrower: we explore the emplacement of various types of knowledge in the processes and procedures of analyzing and processing data. In order to streamline these processes, technical expertise about methods and algorithms is required and knowledge about an exploding number of analytic approaches is very useful. A big deal of this expertise is a tacit knowledge of the experts. However, the identification and subsequent explicit representation of this type of knowledge would be of great value. This paper presents one view on a given topic. It deals with knowledge, used in the process of solving specific tasks of data mining. Our endeavor is to cover various types of knowledge to provide preliminary view on its possible later formalization and representation. The most important types of knowledge that are necessary to successfully solve data analysis tasks are [54]:

- existing different techniques that can be applied in the data analysis process,
- metadata on the input dataset, and
- predictive models, or other types of models.

The whole process is supported by the *domain knowledge* of the given area.

As mentioned earlier, the goal of this chapter is to provide an insight into data scientist's know-how, whereby we deal with expert knowledge, as well as with domain knowledge.

The huge amounts of data are generated daily in almost every area of our life. It is clear, that the value which is hidden in these data can be discovered only if these data are properly analyzed. Our focus is on processing data streams, as due to the contemporary technologies such as IoT, the stream data constitute a majority of currently generated data. From among many data analysis tasks, we concentrate on predictive modeling and optimization. Our group has explored relatively many approaches to solving prediction task. Basically, two approaches to analyze data are used: statistical methods and recently very popular machine learning techniques. We have experience with both types, and we aim to share our experience with them and provide the outlook on the whole area of stream mining, stream forecasting and optimization. In our research, we have verified the proposed methods in the domain of power engineering. Within the scope of this chapter we have selected use cases from the domain of power engineering – also to make feasible to discuss the incorporation of domain knowledge into the model creation, e.g., to include the given constraints into the solution.

In power engineering domain, the traditional grid is changing. It is no more centralized network, where all power customers take energy from the main source. Rather it becomes a distributed network, where the customers will not be just consumers anymore, but they will become prosumers, who consume and produce the energy from renewable energy sources. To effectively use the generated energy, it is important to predict its consumption and production as precisely as possible. Based on these predictions we can then manage the operation of the grid. Here the optimization methods have their firm place. Our goal is therefore to continue the research in the field of optimization methods to offer intelligent smart grid solutions and thus to contribute to an important problem of energy supply.

This chapter is organized as follows: Sect. 2 provides a brief introduction to stream mining in general, in Sect. 3 approaches to stream forecasting are reviewed and our research in this area is mentioned, i.e., our activities towards prediction of energy consumption and production. Based on these outcomes we move on to an important problem of contemporary smart grid – an optimization of its operation. We introduce our efforts in smart grid optimization in Sect. 4 and conclude the chapter in Sect. 5.

2 Stream Mining

The need for efficient processing of large volumes of data resulted in emergence of the stream mining research field. The data streams have similar properties as *big data*: volume, velocity, and variety. Since the data in the stream flow constantly, their volume is essentially infinite. The individual values can come in regular intervals at various speed in one or more simultaneous streams comprising time series. The data can come from diverse sources and change their nature over time due to influence of miscellaneous factors.

Unlike traditional batch mining, stream mining has several restrictions and there are special requirements for data stream processing:

- *accuracy* of stream mining method should not differ significantly from the batch mining method,
- *timeliness* of the stream mining method results should correspond to the interval of data arrival and the results should be provided early enough for decision-making,
- *adaptivity* should be a key part of the stream mining method to cope with the evolving data characteristics in dynamic environments.

Stream Processing and Concept Drift Problem. The learning algorithms in general can be divided into categories based on the amount of data present in the learning process (see Fig. 1). While in offline learning mode a whole training set is needed to create and deploy a model, in online learning mode the data are processed sequentially, and the model is continuously updated with the arriving data. The individual levels of learning differ in the ability to access past data (partial memory) and the frequency of model updates, i.e., batch-by-batch processing in incremental algorithms vs. one-by-one processing in online algorithms [28].

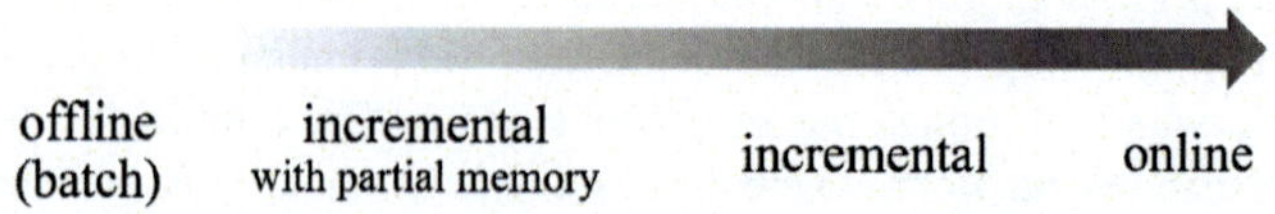

Fig. 1. Types of learning algorithms according to [28].

The incremental and online algorithms can address the last requirement for the stream mining methods – adaptivity, which is closely related to the *concept drift* problem. This problem manifests mainly in the predictive data mining tasks but can be also observed in the results of descriptive tasks if repeated successively over time. "Concept drift means that the statistical properties of the target variable, which the model is trying to predict, change over time in unforeseen ways" [38]. There exist several types of drifts, e.g., abrupt (sudden), gradual, incremental, recurring, etc. [28, 38, 51]. Unfortunately, most of these types are not yet thoroughly formally defined and examined and more studies on their nature are required [35]. Moreover, outliers and noise in data can be easily mistaken for concept drift.

The goal of *adaptive* incremental or online algorithms is to adapt to the changing data stream. Depending on the inclusion of an explicit concept drift detector in the algorithms we differentiate *active* (or informed) and *passive* (or blind) approaches [19, 28, 38]. Ramírez-Gallego et al. [51] identified four main approaches to efficiently tackle the drifting streams: an active approach via *concept drift detectors* and three passive approaches – *sliding widows*, *online learners*, and *ensemble learners*. The passive approaches adapt implicitly over time by continuous learning. The active approaches detect the drift in data stream and react to it only if detected. Lu et al. [38] described three main groups of drift reaction methods in active approaches: *simple retraining* (a new model is retrained and replaces the obsolete model), *ensemble retraining* (a new model is retrained and added to the ensemble of models) and *model adjusting* (a part of the existing model is replaced).

The concept drift problem has been recognized in increasing number of areas, especially in big data community [38]. Lately, several extensive survey papers concerned with this problem [19, 28, 38] or included it as a related issue [35, 51].

3 Stream Forecasting

As mentioned earlier, the stream mining area covers both predictive and descriptive data mining tasks. In our research, we aimed at prediction, i.e. stream forecasting in one particular application domain – power engineering. Prediction of power consumption and production is a typical real-world example of learning in a dynamic and evolving environment [19].

The process of stream forecasting consists of three steps [28]: *predict, diagnose and update* (see Fig. 2). The last two steps are optional since they are part of active adaptive approaches only. These three steps repeat continuously as new samples/batches from the stream arrive.

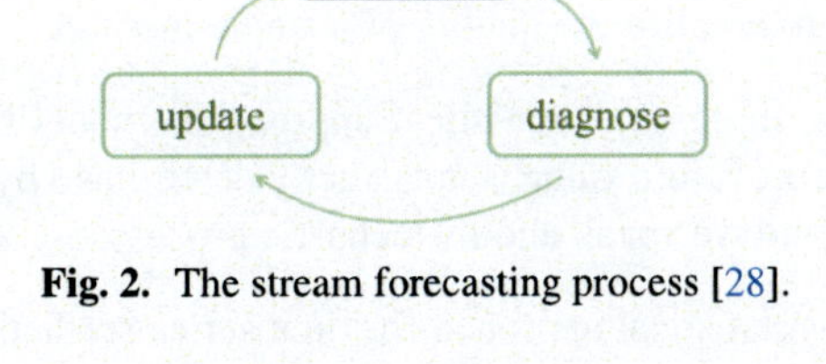

Fig. 2. The stream forecasting process [28].

Open Problems. The most emphasized open problem in the literature is the transformation and optimization of existing offline methods to online or incremental methods [24, 35]. The demand for this kind of methods originate from the ever-growing production of data, i.e., the big data phenomenon, but also from the hardware miniaturization trend that calls for energy efficient methods, which can work with a limited memory space [24]. For successful utilization of offline methods in stream forecasting, the methods require adjustments to achieve *lower computational complexity* and *ability to adapt to concept drifts*.

A significant drop in time and memory complexity can be achieved by removal of redundant data by their approximation [46] and creation of hybrid approaches that combine the prediction methods with methods for transformation of raw time series data to more efficient representations [24]. The complexity reduction and online data pre-processing is suggested to be a mandatory step of the stream mining process [51].

The complexity reduction issue also relates to over-parametrization of mining methods and the need for parameter free data mining approaches [24]. Automated or semi-automated parameter tuning can alleviate this problem, e.g., by utilization of optimization algorithms. This problem is also mentioned in the context of ensemble learning where it is referred to as a problem of self-tuning ensembles [35].

The other way to more efficient mining approaches, is development of methods that are dynamically adaptive to varying nature of data streams, i.e., the concept drift [24]. The concept drift problem requires research of incremental methods that can continuously learn but also forget obsolete knowledge [46]. There is also a lack of research on effective integration of concept drift handling techniques with machine learning methods [38]. The research on concept drift methods is heavily affected by missing benchmarking tools, such as uniform datasets used across papers. The datasets should come from different real-world applications [24, 35, 38] or synthetic data generators [46] and cover various types of concept drifts.

3.1 Power Demand Forecasting

Traditional static time series prediction methods learn the model from the batch of data and are subsequently used without the need to persistently relearn the model. The prediction methods can be divided into two main groups [44]:

1. *statistical*/stochastic methods and
2. *artificial intelligence*/data mining/machine learning methods.

These are often combined into *ensemble models* to exploit the different qualities of multiple methods or to amplify the quality of a single method.

Statistical Methods. In general, statistical approaches model future values based on historical data. Since the future value is not exactly determined by past values, the series is not deterministic, and we speak about stochastic process.

The most popular statistical approach for time series prediction and popular power demand forecasting method is called *Box-Jenkins methodology* [5] and its autoregressive

(AR) and moving average (MA) models and their combinations, such as ARMA, ARIMA [2], ARMAX [36], ARIMAX [71], SARIMA or SARIMAX.

Another popular statistic method for time series analysis and prediction is *exponential smoothing* [9]. Unlike AR and MA models that consider only several past values given by the order of the model and weight them equally, exponential smoothing considers all past values from the beginning and weights them by smoothing factors that decay exponentially. This results to the simple model, where older values have smaller weights than more recent ones. When the time series contains a trend, recursive application of an exponential filter twice or three times results to double and triple exponential smoothing, respectively. Exponential smoothing was successfully used to predict electricity load consumption [61, 62]. In our research, we have extended the double exponential smoothing with concept drift detector based on the domain knowledge – the fact that the daily error of a good power demand forecast should not be higher than 5%. We found out that this incremental approach can significantly save computing resources and provide precise results even when concept drift occurs [64, 65].

Seasonal decomposition of time series by LOESS (STL) [13] decomposes time series into trend, seasonal and remaining components. The decomposition is based on LOESS – a nonparametric regression technique that uses local weighted regression to fit a smooth curve through points in a sequence. STL is robust to outliers. STL was also used to predict monthly electricity consumption [59].

The most utilized family of statistical models used for making continuous predictions is known as *regression analysis*. The dependent variable is predicted based on independent variables. They are known as regressors, exogenous or explanatory variables. In time series analysis, time and calendar variables, such as month, day of a week or hour of a day, are used instead of historical values. Regression models were used to make predictions for power load consumption [11, 68].

AI Methods. The second family of models for time series analysis and prediction solve a regression task by advanced methods of artificial intelligence. The values are modeled based on calendar, time, selected historical data and other exogenous variables.

Artificial neural networks do not require to determine the type of function between input and output in advance. The output predicting function is learned in the form of patterns based on historical data time/calendar variables and/or exogenous variables. Neural networks are very suitable for modeling nonlinearities in data and have a theoretically proven ability to approximate any complex function with arbitrarily chosen accuracy. They have been used extensively to predict electricity consumption since the late 1980s [41]. Their great advantage is that they can naturally model the influence of various exogenous variables on sampling values. On the contrary, the disadvantage is that they belong to so-called black-box models, where the predictions cannot be explained. The most used neural networks are multilayer forward neural networks [34], recurrent neural networks [16] or, recently, deep neural networks [18].

Support vector regression (SVR) [21] is a regression version of the well-known support vector machines (SVM). SVR searches for an appropriate line or hyperplane (in higher dimensions) fitting the data within decision boundaries around the hyperplane in distance given by the tolerable error. Slack variables are usually added into the model to allow approximation in the case that all data cannot lie within the decision boundaries.

Similar to SVM, in the non-linear case, kernel functions are used to transform the data into higher dimensional feature space where data exhibit linearity, without increasing the computational cost. SVR were successfully used for electricity load forecasting [40]. For stream forecasting of power demand, we have extensively studied the online variant of SVR with one-by-one processing and forgetting ability [67].

Regression trees are predictive models based on the decision trees addressing the classification problems. However, the output of regression tree is a continuous value. Usually, it is calculated as an average value of items that ended up in the final leaf of the tree during the tree training. To achieve better prediction accuracy, many strategies how to use multiple regression tree predictors were formed. Regression trees were used for power load forecasting in [33].

The advantages and disadvantages of both statistical and AI methods are summarized in Table 1.

Table 1. Pros and cons of power demand forecasting methods [66].

	Statistical	Artificial intelligence
Pros	• easy interpretability • few parameters to estimate • better univariate models	• minimum statistical or domain knowledge required • ability to model also non-linear relationships between power demand and exogenous variables (e.g., weather) • better multivariate models
Cons	• ability to model only linear relationships • low accuracy in longer term • statistical and domain knowledge required	• difficult interpretability • over-parametrization • heavy computation without optimization

Ensemble Models. The basic idea of ensemble learning is to combine the results of simpler (weak) classifiers or predictors, called base models, to achieve better results. Combined weak methods should be independent of each other and their accuracy should be better than the accuracy of a random model. The three basic approaches are:

- *Bagging* [7] achieves the diversity of basic models by using a bootstrap approach to select a subset of the data on which the model is learned. Random forests [8] represent a certain generalization of bagging. In the process of creating trees, only a randomly selected subset of attributes is used in each division (branching) of the tree.
- *Boosting* approach [53] builds a set of methods sequentially, starting with one model and gradually adding another so that a new model is created based on the errors of the previous base models. Gradient boosting [26, 27], which can be used to optimize any differentiable loss function, is a very popular approach used for regression.
- *Stacking*, unlike previous approaches, where the same types of basic models (e.g., regression trees) are trained, combines the predictions of two or more different basic models (e.g., a regression tree and a support vector regression).

Ben Taieb and Hyndman [60] used gradient boosting to predict electricity consumption. They created semi-parametric additive models and used gradient boosting to estimate the parameters of these models. Papadopoulos and Karakatsanis [48] compared random forests and gradient boosting regression trees with seasonal versions of the ARIMA and ARIMAX prediction models. Ruiz-Abellón et al. [52] compared 4 ensemble methods – bagging, random forest, gradient boosting and so-called condition trees to predict the next 48 samples. The results demonstrated the effectiveness of ensemble learning approaches, especially random forests and gradient boosting. The stacking approach, i.e. combining heterogeneous models to obtain more accurate predictions, has been applied in [10, 20, 47]. We have also proposed an incremental ensemble learning method based on stacking. In the design, we employed the domain knowledge that comprises the knowledge about strong seasonality of the data and the need to consider sudden or gradual concept drifts. The ensemble approach was chosen for its ability to quickly adapt to changes in the distribution of predicted variable and its potential to be more accurate than a single method. Predictions of particular base models can be computed in parallel in order to reduce computation time and to scale up to incoming amount of data which makes the proposed ensemble suitable for big data streams. We selected a wide spectrum of base models and proposed two weighting schemes for combining the base models' predictions – a scheme based on median error of the base models [32] and utilization of biologically inspired optimization algorithms for weighting [4, 31]. We focused on studying the effects of seasonality and concept drifts.

3.2 Power Production Forecasting

The task of dynamic prediction of electricity production from renewable energy sources (RES) is equally important as power demand forecasting and employs similar methods. The production can be unstable, resulting in large actuations of energy, which might cause instability of the grid. Therefore, it is necessary to predict it so that grid operators can plan power generation or effectively regulate the grid to ensure its stability. Photovoltaic (PV) panels obtain energy from solar radiation. Unlike PV, wind energy does not have daily seasonal patterns, instead it depends on local meteorological effects, thermal exchange between ground and atmosphere and general changes in weather patterns. There are three main approaches to prediction of RES energy [1, 55, 69]:

- ***Statistical methods.*** Statistical approaches use only data from the past, which contains information about weather and production of photovoltaic power plant. For a short term, up to 36 h, statistical methods can prove to be a bit accurate.
- ***Physical methods.*** Physical approaches use weather forecasts and technical parameters of photovoltaic power plants or wind turbines, which are necessary for calculation of approximate electricity production.
- ***Ensemble methods.*** Hybrid approaches combine previous approaches to the ensembles to improve prediction. Ensemble or hybrid methods use a combination of different statistical and physical methods with the goal of improving the accuracy.

Artificial intelligence methods are powerful in predicting photovoltaic power production, but their accuracy is highly dependent on their hyperparameter setting. The hyperparameter setting can be done in various ways, either manually or by using algorithms capable of finding and evaluating different hyperparameter settings, such as nature-inspired algorithms. In our most recent research, we proposed the utilization of firefly algorithm to optimize hyperparameters of a SVR model for PV production forecasting [58].

4 Optimization

After a detailed research of forecasting methods, we explored other possibilities of using prediction to increase the efficiency of processes. One of the possible ways to use prediction is in optimization problem, where based on predicted attributes, it is possible to create a plan for a few days ahead. For example, based on the prediction of electricity consumption and production, we will create a plan for the use of batteries to store energy produced by photovoltaics using optimization methods.

4.1 Optimization Process and Open Issues

The aim of the optimization process is to find the best fit result from a set of available feasible solutions that met the conditions of the set model. It is very important to determine the correct model of the problem to be optimized based on domain knowledge. It is necessary to create a formula that describes the real world as much as possible. By setting the values of decision variables, it is possible to maximize or minimize objective functions. Optimization methods are used to solve various types of problems, such as linear, nonlinear, undifferentiated, or dynamic problems. Williamson and Shmoys [72] describe theory and taxonomy of optimization.

The optimization with many objective functions to minimize or maximize is called multi-objective optimization. It is very important to identify objective functions and their constraints, set the limits and conditions correctly [15]. The main problem of multi-objective optimization is that if there is an optimal solution for one function, it is uncertain that it will be optimal for other functions as well. It sets this problem especially when the goals of the objective functions are contradictory. Therefore, the task is to find solutions that are optimal or nearly optimal for each function at once [76]. Pareto optimality represents a solution that is a compromise between all suitable solutions. The problem of pareto optimality is also dealt with in other works, which describe, e.g., the conditions of optimality and the division of the set of solutions [15, 22, 25].

Several open issues need to be addressed when designing a solution for multi-objective optimization. There are several multi-objective optimization review articles that describe currently open issues and challenges [3, 14, 39].

The current research challenges that many authors view from some aspects are:

- *Real time processing and dynamics.* Multi-objective optimization is computationally demanding and therefore no real-time solution is defined for this type of optimization.

It is also difficult to include problem dynamics into the modelling and real-time optimization process.

- *Hierarchy between objective functions.* In two-level problems, when one of the objective functions is subordinate to another, its evaluation is used as a limit in the superior, the computational complexity increases.
- *High dimensionality.* With many objective functions, it is difficult to evaluate the principles of the dominance of found solutions, and therefore most solutions can be evaluated as equally successful.
- *Experimental verification.* Test objective functions are used to evaluate the success of optimization algorithms, but they do not necessarily cover real-world situations. Also, with an increasing number of objective functions the calculation time for evaluation of the objective functions increases [39, 63].
- *Visualization.* The result of evolutionary algorithms for multi-objective optimization is a Pareto-optimal set which contains optimal solutions. To select a suitable solution, it is possible to visualize the results, but with an increasing number of objective functions, it is not trivial.

4.2 Optimization Methods

There are currently several approaches to address optimization. Approaches can be divided into *analytical* (mathematical or exact), *numerical* (approximation or metaheuristics) and their combinations (*hybrid*) [12, 63]. Depending on the complexity of the modelled optimization problem, it is necessary to decide which method to use. If the level of complexity is low enough, it is possible to use analytic methods that guarantee the optimum solution. However, due to the nature of large data and multiple objective functions and with the increasing complexity of the problem, it may be better to use metaheuristics.

Metaheuristics and their exploration and extraction mechanisms provide a reasonably good solution at a reasonably appropriate time [17, 29]. An example of metaheuristic approaches are biologically inspired methods, for example, neural networks, which were created based on the functionality of the human brain, or genetic algorithms, which copy the behavior of evolution in nature. Biologically inspired methods can be divided into three groups: *swarm intelligence*, *evolution-based* and *ecology inspired algorithms*. Bou Ezzeddine et al. [4] described these groups of biologically inspired methods in more detail. These algorithms have been used, for example, in [6], when the mathematical model have been proposed with the aim to optimally manage the smart polygeneration microgrid to minimize daily operational costs and carbon dioxide emission. In the work [45], authors formulated objective functions representing charging and discharging costs, losses, and voltage profile.

The choice of optimization algorithm type depends also on the end user's involvement in the optimization process. In terms of user preferences, according to many authors [15, 22, 75], the methods are divided into four main groups:

- *Methods without knowledge of preferences.* User preferences are not considered, and we want to achieve Pareto's optimal solution, which means a generated neutral compromise. This approach is mainly used when the end-user preferences are not known, and he does not participate in the simulation process.
- *Priori methods.* User preferences are considered and therefore end-user requirements are included in the simulation process. This user information is used for subsequent scalarization of features.
- *Posteriori methods.* The output of the algorithm is a set of solutions from which the end-user can choose the solution that seems to be the best one, in his opinion. Its requirements may be included in the process, but there is no need for its preferences to be included in the input of the algorithms.
- *Interactive methods.* The end user intervenes in the process of selecting solutions for searching in the algorithm solution space so that the selection is directed to its preferred solution. The iteration process is repeated until the user decides to end it.

4.3 Evaluation of Optimization Methods

We can divide the evaluation of the optimization methods based on whether we evaluate *the success of the optimization algorithm* or *the achieved optimization results* [22, 37, 63, 74]. In terms of the success of the optimization algorithm, the result of the multi-objective problems (MOP) should be non-dominated and as close as possible to the Pareto optimal front, just like convergence to the global optimum in single-objective optimization [22, 73]. The most important part of optimization is evaluation of performance of different types of algorithms [14, 37]. The test functions, such as bi-objective Zitzler-Deb-Thiele suite (ZDT) and scalable Deb-Thiele-Lau-mans-Zitzler suite, are used for performance evaluation of multi-objective capabilities in approximating the Pareto front [22]. These function tests non-convexity, multi-modality and non-uniformity of the search space, and discontinuity. Discontinuity of solution usually cause difficulties in solving multi-objective evolutionary algorithms. If the algorithm can solve such test functions, there is a great possibility that it will be able to solve multi-objective problems of the real world.

In terms of evaluating the results of optimization, the most commonly used way to evaluate the success of optimization is to express the values of objective functions. For example, in the domain of power engineering in the process of evaluating the results of optimization, the following indicators can be evaluated: the amount of energy consumed from the mains, the amount of energy consumed from the energy produced by photovoltaic cells, fluctuations and the connection of the curve of energy consumption from the main network, the total price of energy consumed from the mains.

If the stream processing requirements are met, it will be possible to create an online (stream) forecast and then create a real-time optimization plan. Some of the metaheuristic methods can provide multiple solutions, so the decision-maker will be able to choose a suitable solution in time.

4.4 Microgrid Optimization Problems

Microgrids must have a specific management strategy that operates distributed energy resources, trades energy with the main grid and manages the energy load [49, 77]. A hierarchical control structure is used, which is formed of three layers controls. Each layer is used to manage specific actions of the microgrid systems. The primary control layer focuses on real-time control of the microgrid. Its main task is to keep the local voltage and frequency at optimal levels, which is achieved using a technique called droop control [56]. The secondary control layer operates over periods of seconds up to minutes. It aims to compensate for larger deviations in voltage and frequency caused by changes in load or renewable energy production. The tertiary control layer usually operates in matters of hours or days. The main focus of this layer of control is the economic concerns and it manages trading with main grid. This level of control typically involves the collection of information on energy prices, weather forecasts, renewable energy production forecasts and energy load forecasts. Based on these forecasts, a plan for grid trading and generator management is created. This plan is mostly created with economical goal in mind, but it could be even more complex multi-objective one. The schedule can range usually around 24 h.

We assume that the problems of the real world need to be modelled by several specific objective functions that can be optimized. Efficient management of energy consumption, electricity production and storage are a possibility to spend less money on electricity and produce a more accurate amount of energy. There are three main categories that can benefit from optimization in power engineering:

- *Microgrid planning/microgrid sizing.* We can determine the optimal architecture (i.e., the number and type of correct components) due to its expected size and environment with the aim to produce the expected amount of energy consumption.
- *Microgrid operation.* We can adjust the optimal amount of energy produced to the expected consumption and in order to achieve the lowest possible cost.
- *Microgrid control.* We can optimize models describing network status to increase network reliability (i.e., planning, implementation and monitoring of activities of electricity producers).

The application of domain knowledge in the process of prediction and optimization can have a positive impact on more efficient energy management. For example, in case we have knowledge of what kind of electricity consumption it is, we can determine the start-up time of the appliance or limit the amount of consumption when creating a recommendation for the end consumer. Several grid components and their integration are described by many authors [43, 49]. An example of grid components that we should consider when creating a model are:

- electric appliances:
  - fixed – we cannot manipulate the consumption, e.g., security system,
  - flexible – we can manipulate the consumption to optimize the smart grid, there are appliances whose consumption we can limit (e.g., air conditioning or lighting) and those whose consumption we can move (e.g., washing machine, dishwasher),

- distributed energy resources (DER):
  - controllable – we can turn off and on demand or set the power level (e.g., diesel generator),
  - uncontrollable – we cannot control power level, e.g., renewable energy sources (such as solar collectors and wind turbines),
- *devices useful for energy storage*, e.g., energy storage systems with batteries.

Based on the knowledge of the batteries used in the microgrid domain, we can more accurately create a plan for their use. The maximum and minimum battery charge limits are given by its capacity. To prolong battery life, it is possible to change the battery charge and discharge limits settings. Sufyan et al. [57] and Prodan et al. [50] state that battery life decreases with the number of charges and discharges. They define battery life as the number of full charges and discharges until its capacity drops to 80% of its original value. They also state that battery life is also affected by depth of discharge (DOD), which is the opposite of its state of charge (SOC). Weniger et al. [70] mention that in addition to the number of charges and discharges of the battery, its life is also affected by its state of charge. The authors keep SOC batteries in the range of 20% to 80% of its original capacity. To charge the battery to more than 80%, it is necessary to maintain a constant high voltage, which is harmful to the battery in the long run. The limit for batteries is also the maximum amount of battery discharge per hour (e.g., 5 kWh). The worst case of battery use is frequent full discharge cycles. It is optimal to limit the battery charge from 80% to 20% of the original capacity.

Objective functions are identified based on similar domain knowledge. When planning microgrids, the reliability criterion is evaluated, which is usually measured using parameters such as the ratio of *forced outage rates* (FOR), *expected energy not served* (EENS) or *loss of load probability (expectation)* (LOLP (E)) [30]. For example, the LOLP (E) is a criterion that measures how much time (hours) a given energy source will not meet the requirements for supplied energy (i.e., some customers will not receive energy/will be disconnected) during the total number of n hours (1) [23].

$$LOLP = \frac{\sum_{i=0}^{n} Deficit\ load\ time}{n} * 100\% \tag{1}$$

The LOLP of the system can be used as a limitation of the objective function when the value of the LOLP should be lower than the allowable LOLP reliability index.

When optimizing the operation of the microgrid, the goal is to minimize the difference between the energy produced and the electricity consumed. In this case, economic dispatch is used in the short term, and unit commitment in the medium term. For example, in calculating the economic dispatch energy costs are minimized by determining the optimal generator power output, P_k, grid power P_{grid}, and energy storage state of charge SOC_r at each time step i, such objective function is used (2) [42].

$$\min \sum_{i=0}^{n} \left\{ \sum_{k=0}^{g} F(P_k)_i + F\left(P_{grid}\right)_i \right\} + \sum_{r=0}^{s} F(SOC_r)_n \tag{2}$$

There are n time steps $i = 1, 2, 3, \ldots$, and g dispatchable generators and their cost $F(P_k)$. $F\left(P_{grid}\right)$ *is* a time dependent price for either purchasing or selling power and

$F(SOC_r)_n$ represents energy that remains stored in the s energy storage devices at the end of the forecast horizon. In determining this objective function, knowledge of the battery used to store electricity was utilized. Such inclusion in the objective function ensures that solutions which fully deplete the storage are not always preferred. By setting specific optimization goals, it is possible to determine many objective functions and their constrains.

5 Conclusion

The goal of this chapter was to provide a competent view on the recent trends in stream mining with special focus on its application in energy domain. Our team has several years of experience in research of predictive modeling – we have explored many approaches and performed a lot of experiments in searching for effective ways of forecasting power load consumption and production.

The first part of the chapter is therefore devoted to selected methods and techniques of stream forecasting. In our research, we payed a great deal of attention to ensemble learning – a composite model, integrating different individual models, aiming to reduce prediction errors. The potential of ensemble models is known for quite a long time; however, we found a challenge in improving some of its features. The main reason to use ensemble learning was its ability to quickly adapt to changes in the distribution of predicted variable. We proposed an incremental ensemble model for electricity load forecasting based on stacking strategy [32]. We applied biologically inspired algorithms for updating weights in the ensemble [4, 31]. Our interest was to improve the performance of ensemble learning in the presence of different types of concept drift that naturally occur in electricity load measurements. Besides the passively adapting ensemble model, we studied an incremental approach with active adaptation [64, 65] and later an online learning approach [67], which were oriented on effective utilization of computing resources within stream processing. Recently, we focused on application of biologically inspired algorithms on hyperparameter tuning of power production forecasting models [58].

The natural continuation of research in the power load domain is to find effective solutions of controlling the energy supply. Reliable consumption and production predictions are inevitable preconditions of planning process for the next periods. For example, we can schedule the effective usage of batteries to store energy produced by photovoltaics using optimization methods.

In the new concept of electrical network, the term microgrid is becoming very popular. It represents a localized group of electrical sources and loads, that normally operates within the central electrical network, but may also be able to disconnect and to function autonomously. There are several areas of microgrid design, where advanced optimization methods can improve the whole process: propose the optimal architecture (i.e. the number and type of correct components), plan microgrid operation (prepare the optimal amount of produced energy in order to supply expected necessary consumption), establish microgrid control (planning, implementation and monitoring of activities of electricity producers). These problems constitute the challenges for our future work.

Our work was closely associated with identification of knowledge, used in data analysis processes. We considered both types of knowledge – expert and the domain one.

Appointing the relevant know-how of the given area is the first important step before its formalization and utilization. Then, the selection of possible relevant knowledge representations can take place. Among other topics, this will be the subject of our further exploration.

Acknowledgement. This work was partially supported by the Slovak Research and Development Agency under the contract No. APVV-16-0213 for the project "Knowledge-based Approaches for Intelligent Analysis of Big Data"; and the Research and Development Operational Programme for the project "International Centre of Excellence for Research of Intelligent and Secure Information-Communication Technologies and Systems – Phase II", ITMS 313021W404, co-funded by the ERDF.

References

1. Antonanzas, J., et al.: Review of photovoltaic power forecasting. Solar Energy **136**, 78–111 (2016). https://doi.org/10.1016/j.solener.2016.06.069
2. Atique, S., et al.: Forecasting of total daily solar energy generation using ARIMA: a case study. In: 2019 IEEE 9th Annual Computing and Communication Workshop and Conference, CCWC 2019, pp. 114–119. Institute of Electrical and Electronics Engineers Inc. (2019). https://doi.org/10.1109/CCWC.2019.8666481
3. Bechikh, S., et al. (eds.): Recent Advances in Evolutionary Multi-Objective Optimization. Springer, Cham (2017). https://doi.org/10.1007/978-3-319-42978-6
4. Bou Ezzeddine, A., et al.: Using biologically inspired computing to effectively improve prediction models. Int. J. Hybrid Intell. Syst. **13**(2), 99–112 (2016). https://doi.org/10.3233/HIS-160228
5. Box, G.E.P., et al.: Time Series Analysis: Forecasting and Control. Holden-Day, San Francisco (1970)
6. Bracco, S., et al.: A mathematical model for the optimal operation of the University of Genoa smart polygeneration microgrid: evaluation of technical, economic and environmental performance indicators. Energy **64**, 912–922 (2014). https://doi.org/10.1016/j.energy.2013.10.039
7. Breiman, L.: Bagging predictors. Mach. Learn. **24**(2), 123–140 (1996). https://doi.org/10.1023/A:1018054314350
8. Breiman, L.: Random forests. Mach. Learn. **45**(1), 5–32 (2001). https://doi.org/10.1023/A:1010933404324
9. Brown, R.G.: Exponential Smoothing for Predicting Demand. Arthur D. Little Inc., Cambridge (1956)
10. Burger, E.M., Moura, S.J.: Building electricity load forecasting via stacking ensemble learning method with moving horizon optimization (2015)
11. Charlton, N., Singleton, C.: A refined parametric model for short term load forecasting. Int. J. Forecast. **30**(2), 364–368 (2014). https://doi.org/10.1016/j.ijforecast.2013.07.003
12. Cho, J.H., et al.: A survey on modeling and optimizing multi-objective systems. IEEE Commun. Surv. Tutor. **19**(3), 1867–1901 (2017). https://doi.org/10.1109/COMST.2017.2698366
13. Cleveland, R.B., et al.: STL: a seasonal-trend decomposition procedure based on LOESS. J. Off. Stat. **6**(1), 3–73 (1990)
14. Dai, D., et al.: Performance analysis and multi-objective optimization of a stirling engine based on MOPSOCD. Int. J. Therm. Sci. **124**, 399–406 (2018). https://doi.org/10.1016/j.ijthermalsci.2017.10.030

15. Deb, K., et al.: Multi-objective optimization. In: Sengupta, R.N., et al. (eds.) Decision Sciences, pp. 145–184. CRC Press, Boca Raton (2016). https://doi.org/10.1201/9781315183176-4
16. Deihimi, A., Showkati, H.: Application of echo state networks in short-term electric load forecasting. Energy **39**(1), 327–340 (2012). https://doi.org/10.1016/j.energy.2012.01.007
17. Dhaenens, C., Jourdan, L.: Metaheuristics for Big Data. Wiley, Hoboken (2016)
18. Din, G.M.U., Marnerides, A.K.: Short term power load forecasting using deep neural networks. In: 2017 International Conference on Computing, Networking and Communications, ICNC 2017, pp. 594–598. Institute of Electrical and Electronics Engineers Inc. (2017). https://doi.org/10.1109/ICCNC.2017.7876196
19. Ditzler, G., et al.: Learning in nonstationary environments: a survey. IEEE Comput. Intell. Mag. **10**(4), 12–25 (2015). https://doi.org/10.1109/MCI.2015.2471196
20. Divina, F., et al.: Stacking ensemble learning for short-term electricity consumption forecasting. Energies **11**(4), 949 (2018). https://doi.org/10.3390/en11040949
21. Drucker, H., et al.: Support vector regression machines. In: Jordan, M.I., Petsche, T. (eds.) NIPS 1996 Proceedings of the 9th International Conference on Neural Information Processing Systems, pp. 155–161. MIT Press, Cambridge (1996)
22. Elarbi, M., et al.: Multi-objective optimization: classical and evolutionary approaches. In: Bechikh, S., et al. (eds.) Recent Advances in Evolutionary Multi-Objective Optimization. Adaptation, Learning, and Optimization, vol. 20, pp. 1–30. Springer, Cham (2017). https://doi.org/10.1007/978-3-319-42978-6_1
23. Eltamaly, A.M., Mohamed, M.A.: Optimal sizing and designing of hybrid renewable energy systems in smart grid applications. In: Advances in Renewable Energies and Power Technologies, pp. 231–313. Elsevier (2018). https://doi.org/10.1016/B978-0-12-813185-5.00011-5
24. Fakhrazari, A., Vakilzadian, H.: A survey on time series data mining. In: IEEE International Conference on Electro Information Technology, pp. 476–481. IEEE Computer Society (2017). https://doi.org/10.1109/EIT.2017.8053409
25. Feldman, G., et al.: Multi-objective simulation optimization on finite sets: optimal allocation via scalarization. In: 2015 Winter Simulation Conference (WSC), pp. 3610–3621. IEEE (2015). https://doi.org/10.1109/WSC.2015.7408520
26. Friedman, J.H.: Greedy function approximation: a gradient boosting machine. Ann. Stat. **29**(5), 1189–1232 (2001). https://doi.org/10.2307/2699986
27. Friedman, J.H.: Stochastic gradient boosting. Comput. Stat. Data Anal. **38**(4), 367–378 (2002). https://doi.org/10.1016/S0167-9473(01)00065-2
28. Gama, J., et al.: A survey on concept drift adaptation. ACM Comput. Surv. **46**(4), 1–37 (2014). https://doi.org/10.1145/2523813
29. Gendreau, M., Potvin, J.-Y. (eds.): Handbook of Metaheuristics. Springer, Boston (2010). https://doi.org/10.1007/978-1-4419-1665-5
30. Ghorbani, N., et al.: Exchange market algorithm for multi-objective economic emission dispatch and reliability. Procedia Comput. Sci. **120**, 633–640 (2017). https://doi.org/10.1016/j.procs.2017.11.289
31. Grmanová, G., et al.: Application of biologically inspired methods to improve adaptive ensemble learning. In: Pillay, N., et al. (eds.) Advances in Nature and Biologically Inspired Computing (AISC), pp. 235–246. Springer, Cham (2016). https://doi.org/10.1007/978-3-319-27400-3_21
32. Grmanová, G., et al.: Incremental ensemble learning for electricity load forecasting. Acta Polytechnica Hungarica **13**(2), 97–117 (2016). https://doi.org/10.12700/APH.13.2.2016.2.6
33. Hambali, M., et al.: Electric power load forecast using decision tree algorithms. Comput. Inf. Syst. Dev. Inform. Allied Res. J. **7**(4), 29–42 (2016)

34. Kazeminejad, M., et al.: A new short term load forecasting using multilayer perceptron. In: 2nd International Conference on Information and Automation, ICIA 2006, pp. 284–288 (2006). https://doi.org/10.1109/ICINFA.2006.374131
35. Krawczyk, B., et al.: Ensemble learning for data stream analysis: a survey. Inf. Fusion **37**, 132–156 (2017). https://doi.org/10.1016/j.inffus.2017.02.004
36. Li, Y., et al.: An ARMAX model for forecasting the power output of a grid connected photovoltaic system. Renew. Energy **66**, 78–89 (2014). https://doi.org/10.1016/j.renene.2013.11.067
37. Liang, J., et al.: Multimodal multiobjective optimization with differential evolution. Swarm Evol. Comput. **44**, 1028–1059 (2019). https://doi.org/10.1016/j.swevo.2018.10.016
38. Lu, J., et al.: Learning under concept drift: a review. IEEE Trans. Knowl. Data Eng. **31**(12), 2346–2363 (2019). https://doi.org/10.1109/TKDE.2018.2876857
39. von Lücken, C., et al.: A survey on multi-objective evolutionary algorithms for many-objective problems. Comput. Optim. Appl. **58**, 707–756 (2014). https://doi.org/10.1007/s10589-014-9644-1
40. Maldonado, S., et al.: Automatic time series analysis for electric load forecasting via support vector regression. Appl. Soft Comput. J. **83**, 105616 (2019). https://doi.org/10.1016/j.asoc.2019.105616
41. Mansouri, V., Akbari, M.E.: Neural networks in electric load forecasting: a comprehensive survey. J. Artif. Intell. Electr. Eng. **3**(10), 37–50 (2014)
42. McLarty, D., et al.: Dynamic economic dispatch using complementary quadratic programming. Energy **166**, 755–764 (2019). https://doi.org/10.1016/j.energy.2018.10.087
43. Meng, L., et al.: Microgrid supervisory controllers and energy management systems: a literature review. Renew. Sustain. Energy Rev. **60**, 1263–1273 (2016). https://doi.org/10.1016/j.rser.2016.03.003
44. Mishra, N., Jain, Er.A.: Time series data analysis for forecasting – a literature review. Int. J. Mod. Eng. Res. **4**(7), 1–5 (2014)
45. Moradi, M.H., et al.: Improving operation constraints of microgrid using PHEVs and renewable energy sources. Renew. Energy **83**, 543–552 (2015). https://doi.org/10.1016/j.renene.2015.04.064
46. Navaz, A.N., et al.: Real-time data streaming algorithms and processing technologies: a survey. In: 2019 International Conference on Computational Intelligence and Knowledge Economy (ICCIKE), pp. 246–250. IEEE (2019). https://doi.org/10.1109/ICCIKE47802.2019.9004318
47. Neupane, B., et al.: Ensemble prediction model with expert selection for electricity price forecasting. Energies **10**(1), 77 (2017). https://doi.org/10.3390/en10010077
48. Papadopoulos, S., Karakatsanis, I.: Short-term electricity load forecasting using time series and ensemble learning methods. In: 2015 IEEE Power and Energy Conference at Illinois, PECI 2015. Institute of Electrical and Electronics Engineers Inc. (2015). https://doi.org/10.1109/PECI.2015.7064913
49. Parhizi, S., et al.: State of the art in research on microgrids: a review. IEEE Access **3**, 890–925 (2015). https://doi.org/10.1109/ACCESS.2015.2443119
50. Prodan, I., et al.: Fault tolerant predictive control design for reliable microgrid energy management under uncertainties. Energy **91**, 20–34 (2015). https://doi.org/10.1016/j.energy.2015.08.009
51. Ramírez-Gallego, S., et al.: A survey on data preprocessing for data stream mining: current status and future directions. Neurocomputing **239**, 39–57 (2017). https://doi.org/10.1016/j.neucom.2017.01.078
52. Ruiz-Abellón, M., et al.: Load forecasting for a campus university using ensemble methods based on regression trees. Energies **11**(8), 2038 (2018). https://doi.org/10.3390/en11082038
53. Schapire, R.E.: The strength of weak learnability. Mach. Learn. **5**(2), 197–227 (1990). https://doi.org/10.1007/BF00116037

54. Serban, F., et al.: A survey of intelligent assistants for data analysis. ACM Comput. Surv. **45**(3), 1–35 (2013). https://doi.org/10.1145/2480741.2480748
55. Sobri, S., et al.: Solar photovoltaic generation forecasting methods: a review. Energy Convers. Manag. **156**, 459–497 (2018). https://doi.org/10.1016/j.enconman.2017.11.019
56. Soni, U.S., Talwekar, R.H.: Control strategies in droop controller of microgrid: a survey. In: Proceedings - 2018 International Conference on Smart Electric Drives and Power System, ICSEDPS 2018, pp. 239–244. Institute of Electrical and Electronics Engineers Inc. (2018). https://doi.org/10.1109/ICSEDPS.2018.8536054
57. Sufyan, M., et al.: Optimal sizing and energy scheduling of isolated microgrid considering the battery lifetime degradation. PLoS ONE. **14**(2), e0211642 (2019). https://doi.org/10.1371/journal.pone.0211642
58. Sumega, M., et al.: Prediction of photovoltaic power using nature-inspired computing. In: Tan, Y., et al. (eds.) Advances in Swarm Intelligence, ICSI 2020. Lecture Notes in Computer Science, vol. 12145, pp. 25–36. Springer, Cham (2020). https://doi.org/10.1007/978-3-030-53956-6_3
59. Sun, T., et al.: Monthly electricity consumption forecasting method based on X12 and STL decomposition model in an integrated energy system. Math. Probl. Eng. **2019**, 1–6 (2019). https://doi.org/10.1155/2019/9012543
60. Ben Taieb, S., Hyndman, R.J.: A gradient boosting approach to the Kaggle load forecasting competition. Int. J. Forecast. **30**(2), 382–394 (2014). https://doi.org/10.1016/j.ijforecast.2013.07.005
61. Taylor, J.W.: An evaluation of methods for very short-term load forecasting using minute-by-minute British data. Int. J. Forecast. **24**(4), 645–658 (2008). https://doi.org/10.1016/j.ijforecast.2008.07.007
62. Taylor, J.W., McSharry, P.E.: Short-term load forecasting methods: an evaluation based on European data. IEEE Trans. Power Syst. **22**(4), 2213–2219 (2007). https://doi.org/10.1109/TPWRS.2007.907583
63. Trivedi, A., et al.: A survey of multiobjective evolutionary algorithms based on decomposition. IEEE Trans. Evol. Comput. **21**(3), 440–462 (2017). https://doi.org/10.1109/TEVC.2016.2608507
64. Vrablecová, P., et al.: Incremental adaptive time series prediction for power demand forecasting. In: Tan, Y., et al. (eds.) Data Mining and Big Data. LNCS, pp. 83–92. Springer, Cham (2017). https://doi.org/10.1007/978-3-319-61845-6_9
65. Vrablecová, P., et al.: Incremental time series prediction using error-driven informed adaptation. In: Domeniconi, C., et al. (eds.) 2016 IEEE 16th International Conference on Data Mining Workshops (ICDMW), pp. 414–421. IEEE, Piscataway (2016). https://doi.org/10.1109/ICDMW.2016.0065
66. Vrablecová, P.: Predictive analytics on data streams. Inf. Sci. Technol. Bull. ACM Slovak. Chapter **11**(2), 10–18 (2019)
67. Vrablecová, P., et al.: Smart grid load forecasting using online support vector regression. Comput. Electr. Eng. **65**, 102–117 (2018). https://doi.org/10.1016/J.COMPELECENG.2017.07.006
68. Wang, P., et al.: Electric load forecasting with recency effect: a big data approach. Int. J. Forecast. **32**(3), 585–597 (2016). https://doi.org/10.1016/j.ijforecast.2015.09.006
69. Wang, X., et al.: A review of wind power forecasting models. Energy Procedia **12**, 770–778 (2011). https://doi.org/10.1016/j.egypro.2011.10.103
70. Weniger, J., et al.: Sizing of residential PV battery systems. Energy Procedia **46**, 78–87 (2014). https://doi.org/10.1016/j.egypro.2014.01.160
71. Weron, R.: Modeling and Forecasting Electricity Loads and Prices: A Statistical Approach. Wiley, Hoboken (2006)

72. Williamson, D.P., Shmoys, D.B.: The Design of Approximation Algorithms. Cambridge University Press, Cambridge (2011). https://doi.org/10.1017/CBO9780511921735
73. Yu, G., et al.: An a priori knee identification multi-objective evolutionary algorithm based on α - dominance. In: Proceedings of the Genetic and Evolutionary Computation Conference Companion, pp. 241–242. ACM, New York (2019). https://doi.org/10.1145/3319619.3321930
74. Yue, C., et al.: A multiobjective particle swarm optimizer using ring topology for solving multimodal multiobjective problems. IEEE Trans. Evol. Comput. **22**(5), 805–817 (2018). https://doi.org/10.1109/TEVC.2017.2754271
75. Zhang, T., et al.: Scenarios in multi-objective optimisation of process parameters for sustainable machining. Procedia CIRP **26**, 373–378 (2015). https://doi.org/10.1016/j.procir.2014.07.186
76. Zhou, A., et al.: Multiobjective evolutionary algorithms: a survey of the state of the art. Swarm and Evol. Comput. **1**(1), 32–49 (2011). https://doi.org/10.1016/j.swevo.2011.03.001
77. Zia, M.F., et al.: Microgrids energy management systems: a critical review on methods, solutions, and prospects. Appl. Energy **222**, 1033–1055 (2018). https://doi.org/10.1016/J.APENERGY.2018.04.103

Large Astronomical Time Series Pre-processing for Classification Using Artificial Neural Networks

David Andrešič[1(✉)], Petr Šaloun[2], and Bronislava Pečíková[3]

[1] VŠB - Technical University of Ostrava, 17. listopadu 2172/15, 708 00 Ostrava-Poruba, Czech Republic
david.andresic@vsb.cz

[2] Palacky University Olomouc, Krizkovskeho 511/8, 771 47 Olomouc, Czech Republic
petr.saloun@upol.cz

[3] Slovak university of technology in Bratislava, Ilkovičova 2, 842 16 Bratislava, Slovakia
xsuchanovab@stuba.sk

Abstract. During last years, several successful algorithms emerged for classification of time series from various areas of real world. But astronomical time series (a.k.a. light curves containing usually flux or magnitude on one axis and Julian date on the other axis) are a bit more challenging to classify. As they come from multiple observational devices and observatories (designed for e.g. variable stars detection, stellar system analysis or extra-sollar planets discoveries) that are usually located in outer space, they greatly vary in lengths, periods, noisiness and do not have clear borders between classes. Finding periods in these time series is therefore crucial for further research in this area. As these instruments produces huge amount of data (even Petabytes per observing night), we are facing big data issues and the analysis of the data requries an automated solution.

In this chapter, we depict these issues on two publicly available data sets from BRITE and Kepler K2 projects and several well-performing algorithms, such as Convolutional networks, Long Short-term Memory and other Recurrent neural networks etc. We compare these approaches with various data pre-processing methods including e.g. statistic markers, periodograms or Fourier transformation for feature extraction as well as different means of data normalization and balancing. We also compare the affect of various activation functions used in the classification model. At the end, we present our own approach that includes the use of artificial neural networks (Multi-layer perceptron) enhanced by evolutionary algorithm to find and learn the best performing classification model for pre-processed light curves with extracted features. Our approach is able to challenge results of related work that includes these data sets.

Keywords: Time series · Light curves · Artificial neural networks · Multi-layer perceptron · Feature extraction · Classification · Big data · Long short-term memory

J. Paralič et al. (Eds.): DISA 2020, AISC 1281, pp. 117–144, 2021.
https://doi.org/10.1007/978-3-030-63872-6_6

1 Introduction

Time series classification in astronomy can be useful in many areas. In this chapter, we focus on so called light curves: flux or magnitude values measured in some (often irregular) period of time. This can be used to detect variable stars (of many physical classes), extra-solar planets or maybe (in future with more powerful astronomical hardware) extra-solar moons orbiting planets, stellar systems analysis and other physical phenomena.

Since these data comes from various automatized astronomical instruments that produce huge amount of data (even petabytes per night [40]), we are facing big data issues. To handle a classification of this amount of data, an automated solution is crucial. In this chapter we present an overview of current algorithms used for time series classification in general as well as different variations of algorithms based on artificial neural network as a mean of machine learning to test how they perform with astronomical light curves. We also propose a network enhanced by evolutionary algorithm capable to challenge current top results on similar data sets. We also compare the tested approaches with various feature extraction techniques such as statistics markers or e.g. Fourier transformation.

1.1 Types of Stars Variability

In the Universe, most of stars are actually variable in their luminosity. Describing physical reasons of this variability is beyond the scope of this chapter, so we offer at least a brief overview of classes used in our training data:

- *Delta Scuti* - young pulsating stars that are similarly as *cepheids* used as a so called *standard candle* to measure distances between galaxies.
- *Detached Eclipsing Binary* - when both companions in a binary star system have no significant gravitational affect to each other.
- *Semi-Detached/Contact Eclipsing Binary* - when one of the companions in a binary star system is affected by gravitational pull of the other one (but not vice versa) which actually transfers its gas to itself (accretion).
- *Gamma Dor* - young pulsating stars with not very clear physical cause.
- *RR Lyrae ab* - pulsating stars typically with a half of mass of our Sun used as *standard candles*.
- *Other Periodic/Quasi-Periodic*.

In our training data, other stars are labeled as noise. An example of how such time series for these classes looks like is depicted on Fig. 1. On x axis there is always a time (usually some variation of *Julian Date*). On y axis there is usually a flux defined as the total amount of energy that crosses a unit area per unit time[1] (see Eq. 1.1).

$$F = \frac{L}{4\pi r^2} \tag{1}$$

[1] According to *COSMOS - The SAO Encyclopedia of Astronomy*: http://astronomy.swin.edu.au/cosmos/F/Flux.

Where F is the flux at distance r, L is the luminosity of the source star and r is the distance between Earth and the source star.

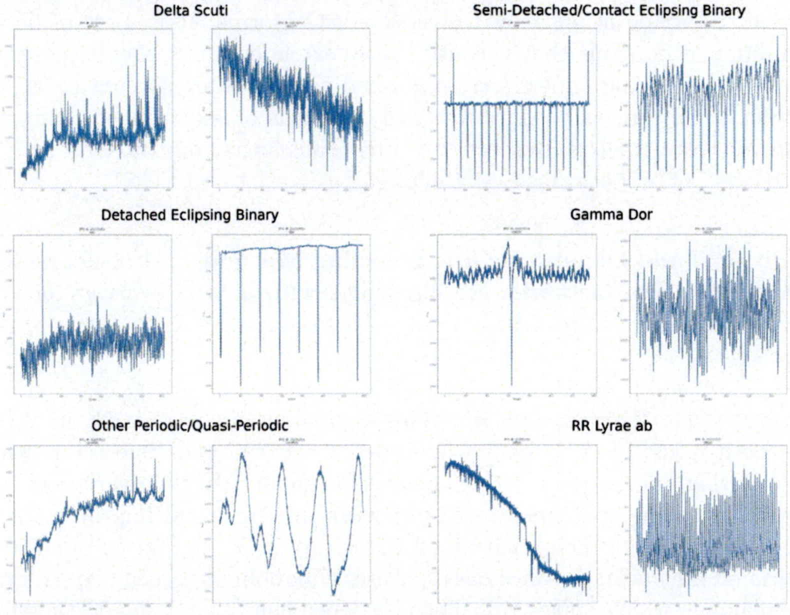

Fig. 1. Example light curves from Kepler K2 data set. Left one from the pair is always with a low confidence in class (under 50%), right ones have high confidence in class (over 90%).

2 State of the Art

Time series classification is a bit specific because the classifier works with sorted data where even in the order of the data a significant markers for some class can be hidden. Most literature on time series classification assumes following [1,23]:

- copious amounts of perfectly aligned atomic patterns can be obtained,
- the patterns are all of equal length,
- every item that we attempt to classify belongs to exactly one of our well-defined classes.

For such time series, the classification using Nearest Neighbour algorithm with the relatively expensive Dynamic Time Warping as a distance measure function usually performs best as a machine learning approach [32]. But these assumptions are challenging in astronomical time series since they greatly vary in lengths, periods, noisiness and are without clear borders between classes. In [26] authors also concludes that Long Short-Term Memory is another state-of-the-art technique for this task.

2.1 Time Series Classification Methods

During last years, several successful machine learning methods emerged for time series classification. They are often bench-marked using UCR Time Series Archive [12] made in 2002 by University of California. It is a set of data sets from different domains that is continuously extended and today it contains 128 data sets. Over 1000 papers were published using this archive until now, but it may not be so conclusive since it heavily depends on experiment configuration details [5]. Nevertheless, authors in [14] concludes that on this data set COTE and HIVE-COTE performs best (followed by ResNet and FCN).

Traditional and Other Machine Learning Methods. Although we focused on artificial neural networks, we also come with a brief overview of current methods and algorithms without them to summarize current state-of-the-art approaches.

Dynamic Time Warping (DTW). This algorithm was introduced in 1978 [35] for speech recognition. It defines a distance between two time series which is a metric that can be used with K-nearest neighbor. Multiple variations of this algorithm has emerged since then, but with only marginal improvement of its original results [38]. The greatest advantage of DTW is that in opposite to a traditional Euclidean distance metric it uses a mapping between two structurally similar points which makes this metric independent [13]. It should therefore be able to identify stars of the same variability class even in case when the stars were not in the same time in the same phase. Another advantage should be the ability to discover a similarity even with different dense of measurements (or speed of change) which means that it could discover a similarity of time series of two variable stars with different gaps between measurements ("sample rate"). It also means that it should be able to discover a similarity even in case of different periods. Authors in [5] compares similar methods of time series classification and it turned out that they did not achieve significantly better results than DTW.

Since we have labeled data, we prefer a supervised learning, which is why we did not consider this method. Beside its use as a metric in kNN, it could also be useful for data pre-processing (e.g. for establishing an average distance of variable/non-variable stars as a one of extracted attributes).

Back of SFA Symbols (BOSS). Introduced in 2015 by Patrick Schäfer [38]. The algorithm is based on a comparison of patterns extracted from time series and contains 3 steps:

1. separation of time series to blocks of same length (sliding windows),
2. transformation of each block using Symbolic Fourier Transformation (SFA),
   (a) a discrete Fourier transform is applied to each block,
   (b) then a symbolic representation of each block is created (SFA word),
3. a construction of BOSS histogram for identifying structural similarities in time series.

BOSS algorithm includes a noise reduction. It is also fast and successful in classification. The author claims that it is able to supersede the best classification algorithms of the time and that one of the modifications could reach the top results on UCR data. But in experiments in [5] it performed a little bit worse.

Transformation of time series could be useful for us as well as a part of data pre-processing. We could also use SFA words themselves or BOSS histograms as inputs for artificial neural network.

Collective of Transform-Based Ensembles (COTE). First introduced in [4] where authors designed a method that collects several time series classifiers in various domains, related transformations to these domains and metrics that evaluates individual classifiers outputs.

Authors in [4] claim that COTE reaches significantly higher accuracy than other known algorithms (those based on artificial neural networks were not covered). This statement can be supported by [5]. COTE was also a part of experiments in [14] where it was compared with 7 other algorithms. Results are depicted on Fig. 2 and COTE took second place with statistically insignificant difference from the winner.

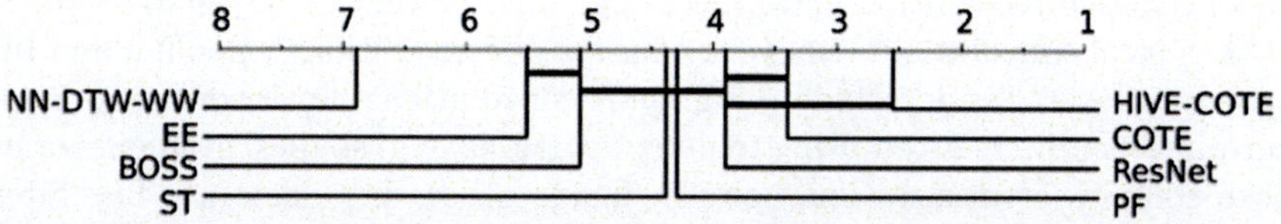

Fig. 2. Critical difference diagram showing pairwise statistical difference comparison of state-of-the-art classifiers on the univariate UCR/UEA time series classification archive. Source: [14].

Methods with Use of Artificial Neural Networks. Methods that utilizes artificial neural networks were the main aim of this work. We bring a brief overview of those that we considered as they are usually performing best for time series from various areas. In [13] authors also proved that artificial neural networks can superseed other time series classification algorithms. We can name for example Convolutional network [28], Fully Convolutional Network [13,26], Residual Network [26], Multi-scalable Convolutional Network [11,13] or Long Short-term Memory (Recurrent) Network [13,26].

Multi-layer Perceptron. Consists of interconnected layers containing artificial neurons. Neurons in two layers are connected by weighted connection. There is no connetion between neurons of the same layer. The output of each neuron consists of a sum of bias and weighted inputs processed by activation function. It is trained by a traditional backpropagation algorithm. More detailed description of MLP can be found in e.g. [33].

In [11] authors attempts to find how each training parameter affects the classifier performance especially in variable stars classification. According to them, the training speed has no significant affect on the accuracy, but size of internal layers has. They also experimented with layers from 4 to 18 neurons and it turned out that those with 4 and 8 neurons performed best. But the main aim of their work was to compare MLP, KNN, SVM and RF in a field of variable stars classification. They attempted to establish a specific classifier for each class and the final class was assigned as a combination of all classifiers outputs. From these, MLP turned out to be somewhere in the middle.

In [5] authors compare algorithms for time series classification that emerged after 2010. It also contains MLP and Naive Bayes (NB), logistic regression, SVM with linear (SVML), quadratic kernel (SVMQ), random forest (RandF) and rotation forest (RotF), 1-nearest neighbor with Euclidean distance (ED) and WEKA C4.5 (C45). From these, MLP achieved better results than DTW, RotF and RandF while it achieved worse results than BN, NB, C45 and logistic regression.

Convolutional Network. Although first convolutional networks were designed for image recognition [15,29,30], today they are commonly used in other domains such as speech recognition and processing [6,42] or time series analysis [16].

A typical convolutional network consists of convolutional, pooling and fully-connected layers (as depicted on Fig. 3). A convolutional layer consists of several convolutional filters attempting to detect patterns (e.g. shapes, gradients or even more complex structures like hairs in image processing domain). The filter is implemented as a matrix used for multiplication of input image. Outputs of convolutional layer are then acumulated in pooling layer made for simplification of inputs by their dimensionality reduction. In the end of the network there is a fully-connected network similar to MLP.

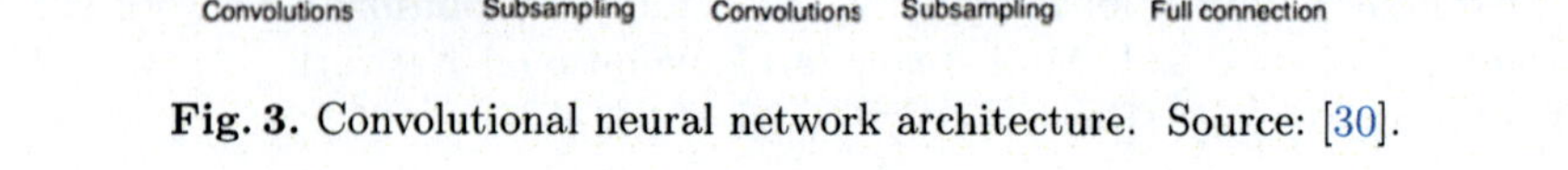

Fig. 3. Convolutional neural network architecture. Source: [30].

In time series domain the convolution can take the place in application of its filter on time series areas. The difference from image processing is that it is applied to one-dimensional inputs. This operation can also be understood as a non-linear transformation of the time series. For example in case of convolution of time series by filter $[\frac{1}{4}, \frac{1}{4}, \frac{1}{4}, \frac{1}{4}]$ we talk about moving average with window of 4. A visualization of convolutional network for time series classification is depicted on Fig. 4.

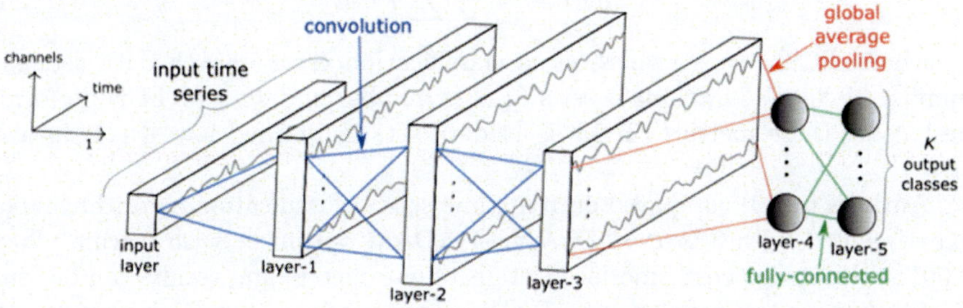

Fig. 4. Fully convolutional neural network architecture for time series classification. Source: [14].

In case of traditional algorithms it is usually necessary to take care for intense data pre-processing [28]. The great advantage of convolutional network is that it requires only minimal modifications of input data for the classifier. This is due to convolutional filters that do the pre-processing on their own. According to [14] the CNN is the most used type of artificial neural network for time series classification, most probably for its robustness a relatively good training speed in comparison to MLP and RNN.

Fully Convolutional Network. The architecture of Fully Convolutional Network (FCN) is similar to CNN. The only difference is that in the last layer, there is (instead of fully connected layer) Global Average Pooling (GAP) layer and that it does not contain local pooling layers so the time series length remains the same during the processing by the entier network. Using GAP significantly reduces the number of parameters for the network. GAP performs dimensionality reduction of tensor with dimensions $h \times w \times d$ to $1 \times 1 \times d$ by reducing each $h \times w$ matrix to arithmetic average of all of its values. Please note that while in image recognition we work with $h, w > 1$, in time series domain one of these parameters is equal to 1 prior the reduction.

The advantage of FCN is that it does not require any significant data pre-processing or features extraction [45] and it can even be used as a mean of pre-processing [26].

FCN achieves great results in domain of semantic segmentation of images [45] and also in field of time series classification [18,26]. The first use of FCN for time series classification is described in [45]. Authors compared FCN, MLP and ResNet with DTW, COTE, BOSS and several other algorithms. They used UCR archive for comparison (with 44 data sets in that time). Data pre-processing contained only z-normalization and for FCN and ResNet training the Adam Optimizer was used. Four metrics were used to evaluate results: arithmetic average of errors accross all datasets, geometric average of errors accross all data sets, number of data sets where the given algorithm performed with lowest error and *Mean Per-Class Error (MPCE)*:

$$MPCE = \frac{1}{K} \sum PCE_k \tag{2}$$

where $PCE_k = \frac{e_k}{c_k}$ where e_k is classification error on kth data set, c_k is number of classes in kth data set and K is number of data sets. FCN performed best using these metrics (in 18 of 44 data sets in reached lowest classification error).

Authors in [14] compared current time series classification algorithms. They experimented with 5-layered FCN with ADAM optimizer with learning factor 0.001 and Entropy cost function. Details of parameters and results can be found in [14], but we can conclude that FCN performs worse that ResNet and better than other 7 classifiers.

FCN was also included in experiments in [41] although this work mostly aims to recurrent neural networks in time series classification domain. FCN with MLP were included into these experiments just to compare the accuracy. It turned out that FCN performs much better than all recurent networks in these experiments (simple recurent network and LSTM).

Residual Network. In general, adding more neurons and layers should increase the approximation accuracy. The problem is that when the network architecture grows, then the backpropagation algorithm faces the vanishing (or exploding) gradient issues. This problem was identified in [20]. Removing this weakness was the major aim of Residual network (ResNet) in order to allow effective network training for deep learning [18]. Another great advantage is (similarly to CNN and FCN) the need of very little data pre-processing [26].

ResNet performs very well in image recognition domain. In [18] authors designed a contest winning network that on ImageNet data set [34] reached only 3.5%. It also turned out that ResNet can be successfully applied to time series classification. In [45] authors experimented with various types of networks and data sets and ResNet achieved lowest classification error in 8 of 44 data sets. In arithmetic average it took 4th place and in geometric average it took 5th place (of 11 algorithms).

Elman's Network. Elman's network is one of the simplest recurrent neural networks (RNN – they store an information about their previous activations in internal memory so the information can spread also among neurons in the same layer; successfully used for sequence data processing [13]). The main difference between MLP and Elman's network is that Elman's network is enhanced with context layer that stores an information about activation of neurons from the previous iteration. These activations then affect the output of the layer using recurrent links. These links exist only between ith neuron of context layer and ith neuron of internal layer and have weight equal to 1. The network is trained using real-time recurrent learning method [46].

In [41] authors attempts to use recurent neural networks in time series classification domain. They compare different neural networks using *UCR Time Series Archive* [12]: MLP, FCN, LSTM and several other types of recurrent networks. They divided recurrent networks into 2 groups: those with dense layer on the output and those with recurrent layer. Their results for recurrent networks are not very conclusive because they did not reach significantly higher accuracy than random classifier. They conclude that according to Wilcoxon signed-rank test:

- replacing recurrent layer by LSTM layer leads to better accuracy,
- adding third layer leads to no significant improvement of accuracy,
- replacing dense layer by recurrent layer has no significant impact to the accuracy,
- increasing number of neurons in one-dimensional recurrent network from 128 to 256 leads to worse classification accuracy.

Long Short-term Memory. The main motivation to create LSTM [21] was to be able effectively model dependencies in large time series and also to eliminate problems with vanishing and exploding gradients [26]. The strength of LSTM comes from regulation of spreading of the activation by gates. These gates regulate input, output and internal state of the LSTM cell. Each LSTM cell consists of 3 gates: input, forget and output. The forget gate decides which information will be removed from internal memory. Input gate filters cell inputs and output gate decides what input values and values from internal layer will be sent to the output of the cell.

LSTM was also a subject of experiments in [41] on *UCR Time Series Archive* [12] where is achieves better accuracy than standard RNN, but worse than FCN. Authors also conclude that increasing the number of neurons in layer from 128 to 256 had no significant affect on classification accuracy. Authors in [19] attempted to classify astronomical time series using LSTM and other methods. Their experiments were evaluated using *Balanced Accuracy*:

$$\frac{\frac{TP}{P} + \frac{TN}{N}}{2} \tag{3}$$

Where TP is a count of correctly classified positive samples and TN is a count of correctly classified negative samples. P is a count of positive samples and N a count of negative samples. Authors were surprised that LSTM performed bad in their experiments using this metrics. They conclude that the *Balanced Accuracy* for LSTM was only 52%.

It seems that noisy time series are quite a challenge for LSTM and RNN in general. Possible solution can be inspired by [17], where the authors used a conversion into a symbolic representation with a self-organizing map.

LSTM Fully Convolutional Neural Network. Introduced in [26] and designed to extend FCNN by LSTM module. Authors state that their solution significantly improves the performance of FCNN with only a small increase of the model. They also state that their solution requires only minimal data pre-processing. They also conclude that their method super-seeds other state-of-the-art methods.

2.2 Other Related Work

In [36] authors performed multi-class classification (instead of our binary one) for the data from original Kepler mission ("K1") that are very similar to ours. Their original results were similarly poor as ours and among others concludes that LSTM is not suitable for Kepler data. They achieved best results with feature extraction. The results are very similar to those described in [7] (again experimented on original Kepler data).

There is also a Kaggle[2] competition aiming on original Kepler data, but although it promises high accuracy, it uses highly unbalanced data set and achieved results are therefore not very informative.

In [22] authors works with a totally different light curves data set, but conclude that feature extraction (in their case a set o 7 statistical markers) is a must-have for astronomical time series. They also work purely with time series without any additional meta-data describing the stars themselves. Based on just these information, they conclude that it is not possible to perform good-performing multi-class classification for most of the variability classes. Using Random Forest, Decision Trees and kNN algorithm they reached up to 70% accuracy which they suspect is caused by an imbalanced data set.

3 Data Sets

In this section we briefly describe data sets we used for experiments with various classification algorithms. From publicly available, real-world data sets such as *All Sky Automated Survey for Supernovae (ASAS-SN)*, *MACHO Project*, *Microvariability & Oscillations of Stars (MOST)* we eventually chose *BRITE* (enhanced by variability data from *GCVS* catalogue) and NASA's *Kepler* (mission *K2*) as they provide data sets in a shape suitable for machine learning (ML).

Real light curves are quite challenging for ML classification. They are very often noisy due to various physical reasons or due to contamination by other signals. In [19] authors for example filtered out those light curves with a large contamination from the neighboring stars or those with total measured flux or a flux yield significantly lower than object's total flux. The sampling rate varies and especially in case of e.g. BRITE data set we can see that whole parts of the time series are missing. Such sparseness is suspected to be one of the reasons why classification using ML does not work very well [7]. Another suspect for ML classification poor results is the density of the data [7], which is why we selected in case of K2 data an original pre-processed data set with 'smoothed' light curves. This data set also removes incorrect values from the light curves caused by instrumentation orientation change performed by on-board thrusters fired during exposure time. This provides more consistent light curves, but on the other hand it brings in another issue which is sparseness. Kepler data are also a bit specific as they contain time series measured in 2 so called cadences

[2] Mystery Planet (99.8%, CNN): https://www.kaggle.com/toregil/mystery-planet-99-8-cnn.

where each have a different sampling rate that affects the density of the data. Some authors overcomes this by simply ignoring them [19]. As we can see, there are many challenges, which is why there are very often used additional meta data like those available in K2 FITS files that comes with photometric data and physical properties of each measured star [7,19]. This - side by side with feature extraction - is the usual ML framework for Kepler data set. In our work, we focused only on raw light curves and labels established by a respected 3rd party without these meta data.

3.1 BRITE

Data from BRITE project - a group of nanosattelites on lower orbit launched in 2013 and 2014. This data set is made of tabular ASCII files containing (among others) Heliocentric Julian Date and Flux (ADU/s). There are 1119 light curves within this data set. According to GVCS catalogue [37], 601 of them are of variable stars, 279 not variable and 239 cannot be identified by cross-matching in GCVS catalogue.

Beside the fact that variability data comes from a 3rd party archive, the greatest disadvantage of this data set is its variability in sampling rate (in some cases, intervals between samples are less than 1 ms on one side and approximately 60 days on the other side). Another disadvantage is that number of samples in each light curve varies (some light curves have less than 10 measurements, while others have tens of thousands of measurements). The average length of light curve is 12642 of samples [1].

General Catalogue of Variable Stars (GCVS). GCVS [37] is a list of known variable stars (it collects this knowledge from research in astronomy domain). Its first version contained 10820 stars and was released in 1948. Since then it was updated several times and today in contains 52011 variable stars (version 5.1 released in 2015). This catalogue allows a cross identification of stars based on their IDs in various catalogues.

Since BRITE data set does not come with an information about star variability, we used this catalogue to add an information about variability by cross matching its ID. By this we were able to obtain a label (7 classes of star variability) for 1119 light curves.

3.2 Kepler K2

NASA's mission to search Earth-like extra-solar planets in our Galaxy. Data from K2 mission that are subject of this work are publicly available [2,3] and well documented [10,25]. In this work we are interested in Kepler K2 light curves containing flux of individual objects in time. For these data, Kepler K2 also provides official catalogue of confirmed variable objects that we can utilize. Based on sampling frequency, we distinguish 2 cadency groups: long with 1765.5 s (29.4 min) and short with 58.89 s. On each Thursday, more than 160000 objects with long cadency and 512 objects with short cadency were measured and archived. The minimal length of measurement was 1/4 of year for long cadency and 1 month for short cadency (with exception of Q4 where module 3 objects were lost due to hardware failure). Light curve file is in a form of time series where all undefined values are represented as NaN (not a number). As a result, we can obtain about 40000 light curves with length up to 1300 measurements [1].

As mentioned before, we used original, but corrected K2 data that excludes observations during thruster firings [44]. The resulting data are more smooth and without irrelevant samples (although with some sparseness). The difference between raw and corrected data is depicted on Fig. 5.

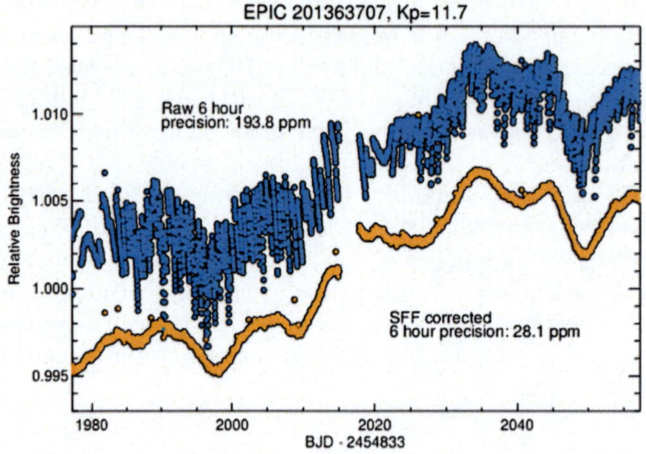

Fig. 5. An example of the uncorrected fluxes from K2 (blue) and the corrected K2SFF version (orange). Source: [44]. (Color figure online)

Similarity with Kepler "K1" Data. Kepler mission "K2" followed the original NASA's Kepler mission with the same purpose after the instrument stabilization failure. For this reason, these data sets are very close to each other in terms of the content. Original Kepler data also contains light curves containing flux values with two cadences: long with 29.4 min and short with 58.89 s sampling frequency.

Currently, more research seems to be done on original ("K1") Kepler data, but due to the data similarity, we include their results in our work as well. In [19] authors performed multi-class classification but with poor results (balanced accuracy 52%). They confirm that LSTM does not perform very well for Kepler data (they discuss that it was either due to the limited positive sample size within our data or the sparseness and/or noisiness of real data) and they achieved best results with use of significant attributes extraction (with a balanced accuracy 74.7%). This conclusion was basically confirmed by experiments performed in

[7] where authors performed also multi-class classification of 150000 objects into 14 variable star classes reaching up to 65–70% accuracy. Authors also mention previous research on Kepler data achieving up to 55% accuracy.

4 Artificial Neural Networks Approaches

We decided to compare several methods of time series classification that utilizes artificial neural networks. Our primary goal is to find the best method that will classify time series (light curves) at least binary in a sense of object variability: the light curve contains some period (and is therefore of variable object) or whether there is no period found (an object is not variable) [1] as much as the data will allow us. We started with following approaches:

- multi-layer perceptron classification with own activation function,
- recurrent neural network of type LSTM,
- multi-layer perceptron with own activation function in combination with time series pre-processing using Fourier transformation,
- recurrent neural network of type LSTM in combination with time series pre-processing using Fourier transformation,
- multi-layer perceptron classification with own activation function in combination with some method of significant attribute extraction,
- recurrent neural network of type LSTM in combination with some method of significant attribute extraction (a.k.a. feature extraction),
- convolutional neural network with sigmoid activation function,
- other, not so successfull (in terms of our results) artificial neural networks such as fully-convolutional neural networks, MCDCNN and ResNet.

After these we attempted to evolutionary engineer ANN specifically for our binary classification task and establish a framework that can match the classification accuracy of related work.

4.1 Data Pre-processing

We pre-processed both BRITE and Kepler K2 data sets and transformed the raw data (light curves with flux) into a common form digestible by the ANN:

- Balancing data sets so it contains same number of variables and not variables.
- Cutting light curves in order to equal their length.
- Mix the data.
- Generate periodogram.
- Perform Fourier transformation.
- Significant attributes extraction in order to reduce dimensionality of time series data (different techniques).
- Data normalization into interval relevant to selected activation function.
- Splitting the data set to training and test set.

Concrete details of these pre-processing approaches can be found later in description of individual experiments as we experimented with various settings.

Significant Attributes Extraction and Visualization. During our experiments described later we discovered that both original data sets may not provide clear examples of time series of variable and non-variable stars. This led to a poor accuracy and we were therefore looking for a way how to distinguish these time series by means of significant attributes extraction. We tested the usability of extracted attributes by visualization using Sammon mapping [36]. Sammon mapping attempts to find a low-dimensionality representation of objects in high-dimensional space with as much respect to their original geometric distances as possible. We used it to convert extracted significant attributes to 2D and visualize, hoping to see clear clusters with variable and non-variable time series. Such set of attributes could be then used for further classification using ANN.

5 Experiments and Results

As we achieved poor initial results with BRITE data set (as described in Sect. 5.1, not all experiments covers it. We were looking at results with 10 following activation functions: exponential, sigmoid, hyperbolic tangent, relu, elu, selu, soft plus, softsign, harp sigmoid, linear. Experiments were performed with artificial neural network containing 3 hidden layers: 16 neurons in input, 34 neurons in first hidden, 16 in second hidden and 64 in third hidden layer [1].

5.1 BRITE Data Set

We had started with a more problematic BRITE data set. We attempted to classify time series by several ways, but eventually with poor results.

Multi-layer Perceptron. The sigmoid activation function was used in the output layer. Training was stopped after 3000 epochs. The learning rate was set to 0.005. For each activation function we trained the MLP 10-times and based on validation data we selected the best model. Its accuracy was then tested on testing data. Results can be seen in Table 1.

Bad results are probably caused by a small data set. Only 538 light curves came out from pre-processing, these were then divided to a training and test set in 70:30 ratio, 10% of training set was used for validation. For the training, only 338 light curves remained. Another issue was the irregular interval between individual measurements within the time series. Based on these results and results with LSTM, we decided to continue only with Kepler K2 data set.

Long Short-Term Memory. In this case, the process was a bit different. We used 900 raw time series (as LSTM is supposed to handle it) cross-matched with GCVS catalogue. We divided them into training and test set in 70:30 ratio. Each time series had up to 66500 measurements ("feature vector"). Shorter time series were padded with −1 to this length and all data normalized. With these settings, we achieved the accuracy 60%.

Table 1. Results of MLP classification for BRITE data set.

Act. function	Precision	Recall	Accuracy
Exponential	0	N/A	0.64
Sigmoid	0	N/A	0.64
Hyperb. tan	0.03	0.18	0.59
Relu	0	0.60	0.64
Elu	0	N/A	0.64
Selu	0	N/A	0.64
Soft plus	0	N/A	0.64
Soft sign	0	0	0.62
Harp sigmoid	0	N/A	0.64
Linear	0	0	0.61

5.2 Kepler K2 Data Set

After attempts with BRITE data set, we switched to a more promising Kepler K2 data set. Configuration was same as in case of BRITE.

Multi-layer Perceptron. The results with the same configuration as in case of BRITE can be seen in Table 2. Unfortunately, there is just minimal improvement. The best activation function turned out to be Selu that achieved accuracy 0.66. Recall is also interesting metric because it is not such an issue if some non-variable object is classified as variable but it is important to minimize the number of undetected variables. From this point of view, the Elu function performed best. Accuracy and loss function of best models is depicted on Fig. 6 and 7.

Table 2. Results of MLP classification for Kepler K2 data set.

Act. function	Precision	Recall	Accuracy
Sigmoid	0.98	0.53	0.56
Hyperbolic tangent	0.72	0.61	0.64
Relu	0.67	0.60	0.61
Elu	0.64	0.63	0.63
Selu	0.78	0.62	0.66
Soft plus	0.82	0.58	0.62
Soft sign	0.64	0.58	0.60
Harp sigmoid	0.98	0.49	0.48
Linear	0.73	0.61	0.63

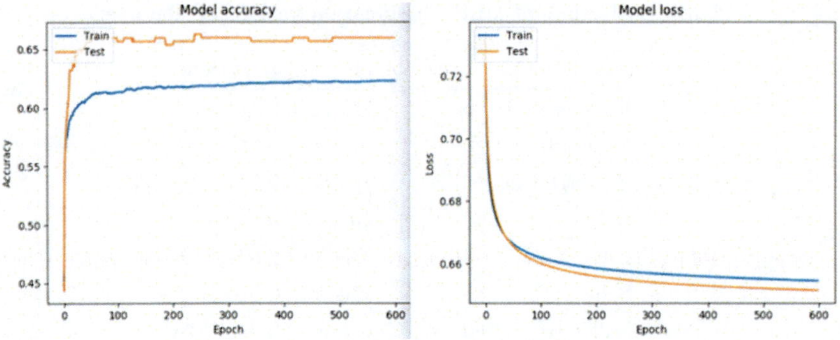

Fig. 6. Results of MLP training with Elu activation function on K2 data set. Source: [1].

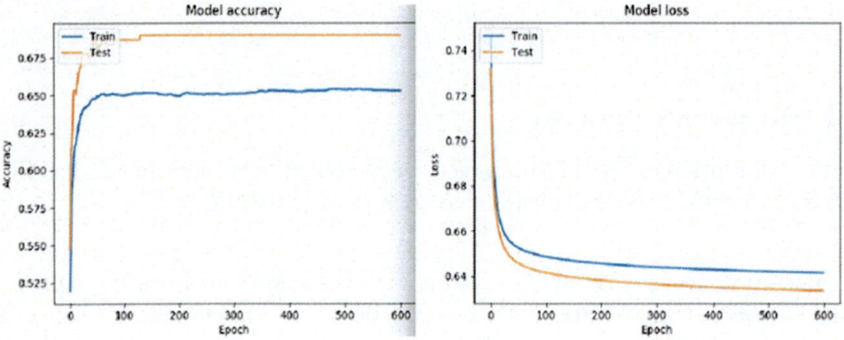

Fig. 7. Results of MLP training with Selu activation function on K2 data set. Source: [1].

Then we attempted to improve the accuracy by generating so-called periodograms created by conventional statistical analysis. The classifier then attempted to classify these periodograms instead of light curves. Results can be seen in Table 3 and shows no significant improvement of accuracy. Nevertheless, hyperbolic tangens performed best. We also attempted to establish some custom non-linear activation functions listed in Table 4 to see if activation functions can have any significant impact on classification accuracy (raw light curves were used). As last experiment with MLP, we attempted to use our own activation functions with Kepler K2 light curves processed by Fourier transformation. Results are listed in Table 5. We achieved similar results with Cosine transformation.

Convolutional Network. We decided to compare CNN with MLP. All data from K2 data set were normalized by min-max normalization, we run 5000 epochs with learning rate 0.005 and following network configuration: two convolutional layers with sigmoid act. function, window size of 7 and 6 (or 12) filters, each followed by pooling layer and with output layer with sigmoid activation function. Results are in Table 6 and on Fig. 8. We can see no significant improvement.

Table 3. MLP classification for K2 data converted to periodograms.

Act. function	Precision	Recall	Accuracy
Sigmoid	1.00	0.49	0.49
Hyperbolic tangent	0.51	0.71	0.66
Relu	0.54	0.70	0.65
Elu	0.63	0.65	0.65
Selu	0.65	0.61	0.62
Soft plus	0.80	0.61	0.65
Soft sign	0.56	0.67	0.64
Harp sigmoid	0.00	0.00	0.50
Linear	0.61	0.65	0.64

Table 4. MLP classification for K2 data using own act. functions.

Act. function	Precision	Recall	Accuracy
tanh(0.1*x)	0	0	0.45
tanh(0.3*x)	0.05	0.53	0.45
tanh(0.5*x)	0.52	0.66	0.59
tanh(x)	0.72	0.68	0.66
tanh(1.5*x)	0.68	0.68	0.65
tanh(2*x)	0.71	0.67	0.65

Table 5. MLP classification for K2, own act. functions, FT.

Act. function	Precision	Recall	Accuracy
tanh(1.1*x)	0.68	0.64	0.64
tanh(1.4*x)	0.75	0.65	0.67
tanh(1.5*x)	0.74	0.65	0.66
tanh(1.7*x)	0.68	0.61	0.61
tanh(2*x)	0.72	0.63	0.63

Table 6. Results of CNN classification for Kepler K2 data with different count of light curves and measurements in each time series (cutted to this length).

Light curves	Length	Precision	Acc.	Recall
500	800	0.65	0.65	0.65
7502	1300	0.65	0.64	0.64
500	400	0.64	0.62	0.63

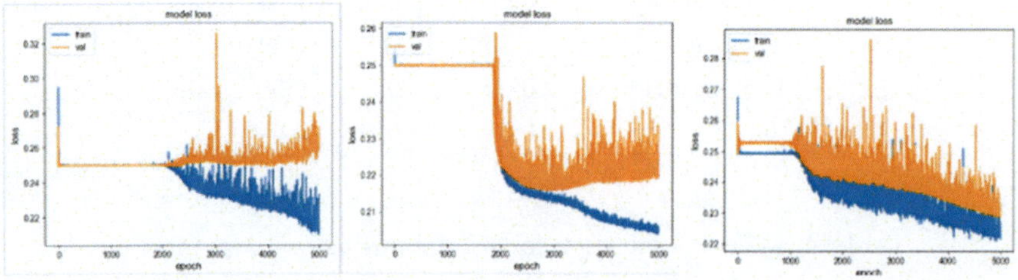

Fig. 8. CNN experiment loss function chart during training phase (from left: eperiment #1, #2, #3) [1].

Other Artifical Neural Networks. We have been experimenting with several other ANNs including ResNet, Fully-convolutional network, MCDCNN and other configurations of MLP and CNN with even less success than was described above. Their results are therefore omitted from this paper.

Other Significant Attributes Extraction Methods. As mentioned before, we attempted to extract most significant attributes from K2 data set in order to distinguish variable and non-variable objects. To verify this, Sammon projection was used (see Fig. 9 and 10). The experiment confirmed that the original data does not contain clusters and we need to focus on domain-specific details of the data.

MLP: Improvement of Accuracy. We eventually focused on the most-promising MLP with use of feature extraction and certain domain knowledge. We used Kepler K2 variability metadata that also provides a confidence level of each class label (in a form of probabilistic distribution over all possible classes). The histogram of such label confidence is depicted on Fig. 11. For the training purposes, we fine-selected only those time series that has confidence level over 80% and thus are more "clean" for the feature extraction (in other cases there is a significant probability of a bad label) – an experimentally established set of statistical markers (calculated using *tsfresh*[3] framework):

- *Absolute sum of changes* of the time series x: $\sum_{i=1}^{n-1} |x_{i+1} - x_i|$.

[3] tsfresh: https://tsfresh.readthedocs.io.

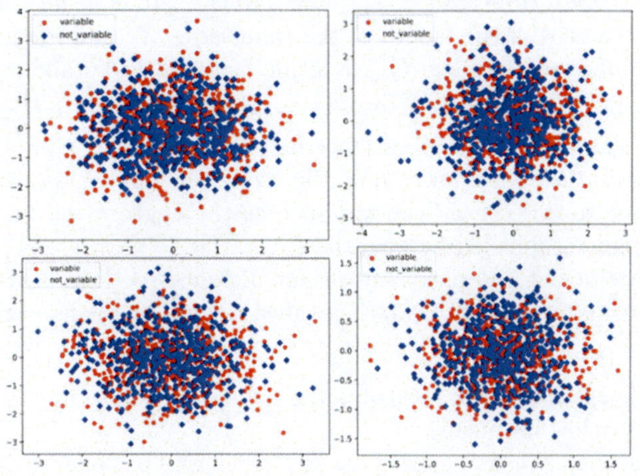

Fig. 9. Sammon projection, random init.: 500 epochs, different time series length (1200–1225 measurements), extracted attributes or just FT or min-max norm [1].

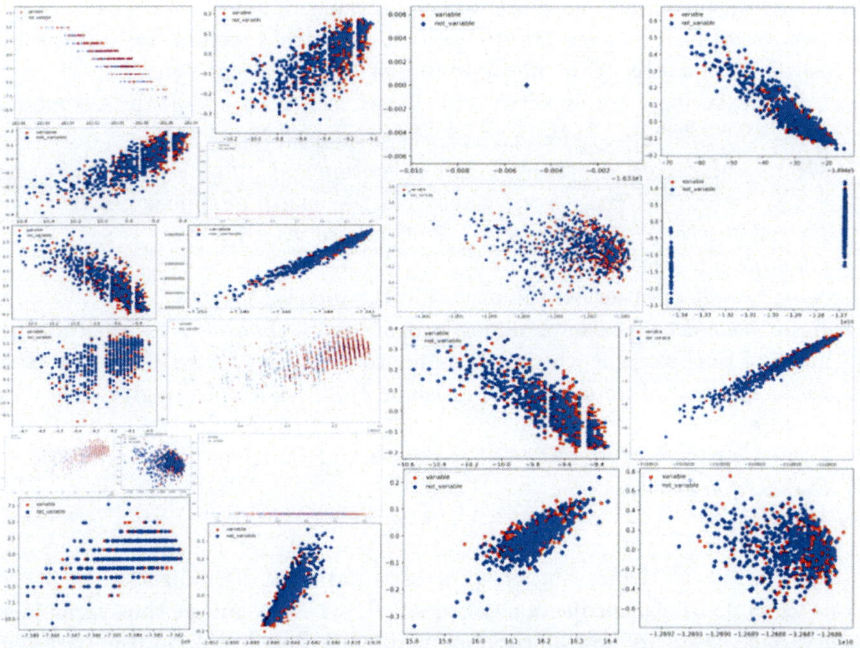

Fig. 10. Sammon mapping: 500 epochs, time series length: 1200–1225. Attr.: stand. dev., variance, min, max, mean, sum val., median, abs. sum of changes, agg. autocorr., arithmetic coeff., binned entropy, energy ratio, agg. FFT val., first loc. of min/max, mult. max values, index mass quant., linear trend etc. Or processed by FT or min-max normalization [1].

- *Aggregated auto-correlation*: $f_{agg} = (R(1), ..., R(m))$ for $m = max(n, maxlag)$ where n is the length of the time series X, $maxlag$ is the maximal number of lags to consider, f_{agg} is mean, variance and standard deviation in our case and $R(l)$ is the autocorrelation for lag l: $R(l) = \frac{1}{(n-l)\sigma^2} \sum_{t=1}^{n-l}(X_t - \mu)(X_{t+l} - \mu)$ with σ^2 being variance and μ mean.
- *Change quantiles* with lower quantile being 0.5, higher quantile being 0.7, using absolute differences and variance as the aggregation function applied to a corridor established by quantiles.
- *CID* - an efficient complexity-invariant distance for time series x attempting to estimate its complexity specified by more peaks, valleys etc. [8]: $\sqrt{\sum_{i=0}^{n-2lag}(x_i - x_{i+1})^2}$.
- *Count above/below mean* returning the number of values in time series x that are above/below its mean.
- *Energy ratio by chunks* - sum of squares of chunk i out of N chunks expressed as a ratio with the sum of squares over the whole series (we used 10 chunks).
- *Fast Fourier coefficients* $A_k = \sum_{m=0}^{n-1} a_m e^{-2\pi i \frac{mk}{n}}$ for $k = 0, ..., n-1$ where A_k is the kth Fourier (complex) coefficient, n is the length of the time series and a_m is the mth value of time series. We used first 3 Fourier coefficients: their real, imaginary, absolute and angle value.
- *Aggregated Fast Fourier Transformation* - spectral centroid (mean), variance, skew, and kurtosis of the absolute fourier transform spectrum.
- *C3* measuring non-linearity in time series x [39] by computing $\frac{1}{n-2lag} \sum_{i=0}^{n-2lag} x_{i+2lag}^2 \cdot x_{i+lag} \cdot x_i$ where we used $lag = 350$.
- *Mean value of a central approximation of the second derivative*: $\frac{1}{n} \sum_{i=1}^{n-1} \frac{1}{2}(x_{i+2} - 2x_{i+1} + x_i)$ where n is the length of time series x.
- *Partial autocorrelation* at lag $k = 2$ of time series x [9].
- *Quantile* $q = 0.5$.
- *Range count* of observed values within the interval $[0, 100)$.
- *Ratio beyond* $r\sigma$ - ratio of values that are more than $r \cdot \sigma(x)$ away from the mean of time series x where σ is standard deviation a $r = 100$ in our case.
- *Ratio of count of unique values to count of all values* of the give time series.
- *Skewness.*
- *Power spectrum of the different frequencies* of the given time series.
- *Standard deviation.*
- *Variance.*

On Fig. 12 we depict our pre-processed data set. Visualization in 3D was done using non-linear projection called t-SNE [31]. We can see that variable and non-variable stars are now much more distinguishable. In the bottom part where most of non-variable stars are located, we can see a significant number variable stars as well which suggests that there will be a need of rather higher number of layers and neurons in them in order to "bend" the space around them and separate them in high-dimensional space.

We also introduced batch normalization [24] for each hidden layer. The best experimentally found performing network contained 3 dense layers (64, 128 and

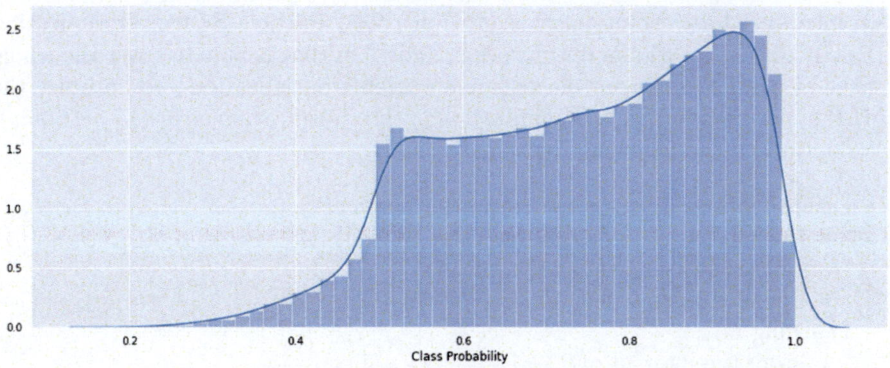

Fig. 11. Histogram of class probabilities (label confidence) for Kepler K2 data. We performed our best experiments with light curves having class probability over 80% and 97% so we can see how representative our sample of data was.

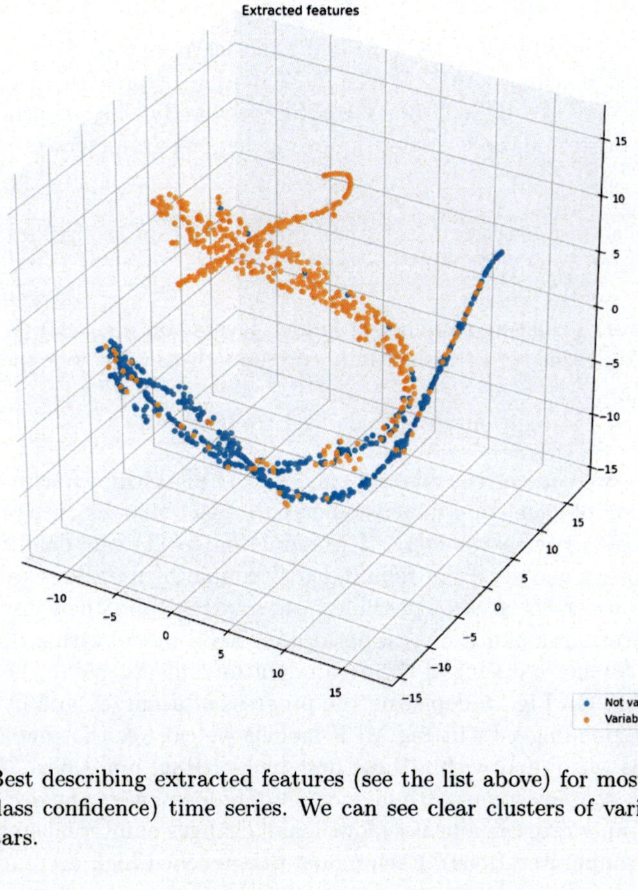

Fig. 12. Best describing extracted features (see the list above) for most reliable (in a sense of class confidence) time series. We can see clear clusters of variable and non-variable stars.

256 neurons) and reached accuracy over 70%. To push the accuracy even further, we introduced an evolutionary algorithm called Particle Swarm Optimization (PSO) [27] - a traditional optimization algorithm that is able to cover the whole space of possible solutions on its random agents initialization - to optimize the MLP hyper-parameters by minimization of the classification error (1 - *accuracy*). We optimized the following hyper-parameters:

- Decimal places of extracted statistical markers
  - Encoded as rounded integer value.
- Normalization type (min/max, mean)
  - Encoded as rounded value of 0 or 1.
- Learning rate
  - From the interval of 0 to 1.
- MLP network structure (number of hidden layers and neurons).
  - Up to 10 hidden layers having different combinations of 8, 16, 32, 64, 128, 256, 512 and 1024 neurons.
  - 43757 possible structures in total.
  - Encoded as rounded integer value that represents one of layers combinations.

The whole process is depicted on Fig. 13 and learning took 5 days on modern CUDA-enabled machine (AMD Ryzen 7 2700 with 8 cores/16 threads, 32 GB RAM, nVidia GeForce RTX 2070 8 GB, SSD drive). By this, we achieved accuracy 98.55% on test data set using following hyper-parameters established by PSO:

- Decimal places of extracted statistical markers: 7 (not so significant).
- Normalization type (min/max, mean): mean.
- Learning rate: 0.1196596.
- MLP network structure (number of hidden layers and neurons): 8, 16, 16, 32, 32, 32, 1024, 1024, 1024 (this actually corresponds to what we expected based on Fig. 12).
- Achieved by 814 light curves in balanced training set.

On Fig. 14 we can see the convergence of PSO algorithm. Each point represents a position of agent in 4-dimensional space (each dimension represents one optimized MLP hyper-parameter). The reduction to 2D was done using PCA algorithm. As we can see PSO required just couple of iterations (about 20 or 30) to find the correct area of possible best solutions and then just spent the remaining iterations looking for best performing MLP model within this area. We suspect that further reducing of the area might be difficult due to the stochastic nature of MLP. On Fig. 15 depicting the progress of accuracy and hyperparameters of corresponding best fitting MLP models we can see that one of the best fitting models was found within these first tens of PSO iterations. We can also see that high accuracy comes with high count of hidden layers and decimal places while the learning rate remains very low (small changes of internal states of MLP during the training iterations). Each model that reached high accuracy also had

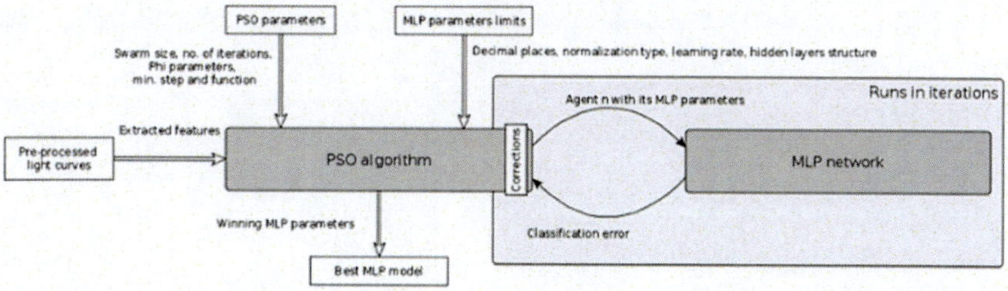

Fig. 13. PSO over MLP. We start with feeding PSO with initial swarm parameters and limits for MLP parameters that are optimized by PSO. In each iteration of PSO, n agents runs MLP classification with optimized parameters producing the MLP classification error. Based on this error, corrections to MLP parameters of the given agent are made for the next iteration.

similar structure of hidden layers starting from 8 neurons at the beginning layers and ending with 1024 neurons among last layers.

We successfully reproduced the experiment with basically same results and accuracy. It seems that 98.55% accuracy is the current top for our approach to binary classification of Kepler K2 light curves. We are also aware of a bias introduced by optimizing against test accuracy, which is why we watched close the validation accuracy as well as precision and recall where all these metrics were over 90% too. We also tested our approach with data having label with lower confidence (over 80%) and achieved 82.59% accuracy after only 15 iterations (according to Fig. 14 and Fig. 15 there is only minimal improvement possible as PSO finds good models very fast). The confusion matrix is depicted in Table 7.

Table 7. Confusion matrix for our experiment with MLP optimized by PSO algorithm for light curves with class probability (label confidence) of 97%.

		Actual class	
		Variable	Non-variable
Predicted class	Variable	73	1
	Non-variable	3	61

Results Summary. Similarly to [19] we achieved poor results with LSTM and best results using feature extraction. Although in [19] and [7] authors worked with original Kepler mission data and multi-class classification (instead of our binary one), we were able to reach higher accuracy on a very similar data Kepler K2 data set and binary classification. It seems that the key lies in data preprocessing and selection of correct training data as the confidence level for the label data is often deeply under 50%. Comparison with Kaggle competition

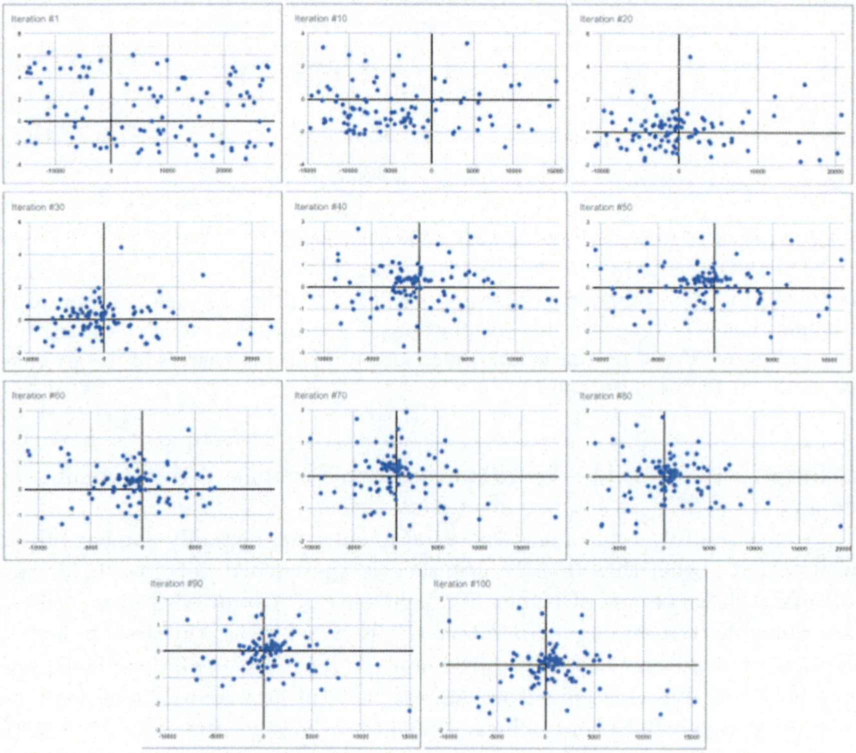

Fig. 14. PSO agents convergence (per tens of iterations from top left to bottom right).

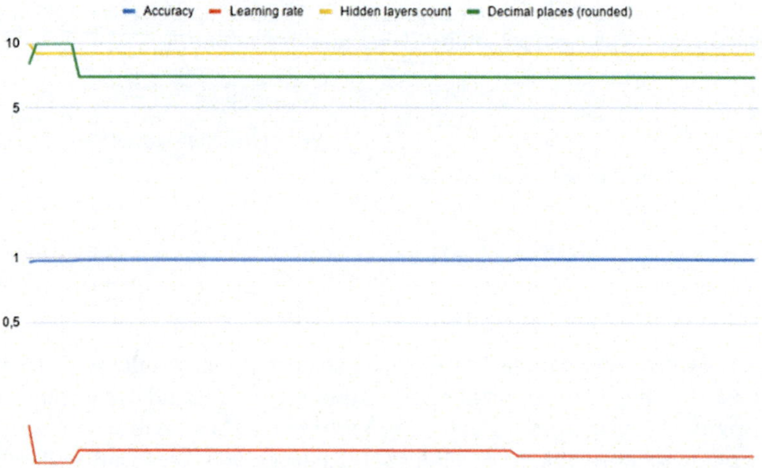

Fig. 15. Best fitting MLP models over PSO iterations and their hyper-parameters. We can see that model with reasonably high accuracy was found during first iterations.

results is also problematic not just for the fact that it also uses the original Kepler mission data, but also for using a highly imbalanced data set.

6 Conclusions

Astronomical time series (a.k.a. light curves) are very specific in comparison to other ones - they are noisy, variable in length and observational intervals and with overlapping classes. Traditional methods for their classification seems to not perform very well with them. Significant attributes extraction that includes various statistical markers and FFT coefficients seems to be necessary to achieve good results and the classification accuracy can be highly improved by optimizing the hyper-parameters with evolutionary algorithm. It also seems that successful classification model must contain a high number of hidden layers ("deep learning"), low training interval (in case of MLP that is also a good choice for classification based on extracted features) and data with high precision. Also a structure of hidden layers seems to have an impact on classification accuracy as we observed small number of neurons in layers at the beginning of the network and high number of neurons among the last layers of the network.

Our approach is time consuming, but good classifiers can be found very quickly during first iterations and then they are just fine-tuned. In the end it can reach high accuracy confirmed by validation and test set to prevent overfitting. So far we are not aware of a related work reaching similar accuracy on Kepler data set. Similar work usually performs multi-class classification with help of photometric metadata as additional features while we rely just on the time series themselves (although we select only those with high confidence in label). No other work using Kepler data that we are aware of also mentions what kind of or how much data they used (that is, whether the data set was balanced, what was the number of samples used or what was the minimum or mean of the label confidence).

We expect that by introducing photometric meta-data we can lower the confidence in label while keeping similar accuracy in the future. We would also like to confirm this approach on different astronomical data sets and possibly on multi-class classification (as samples of individual classes may differ too much from each other to be united as one 'variable' class). The speed is also challenging although we can see that good classifiers are found quite quickly. We are aware of papers that attempts to speed up the convergence of hyperparameters optimization using e.g. Decision trees or Bayesian optimization. A nice overview of optimizing (deep) neural networks also offers e.g. [43]. Other possibility for binary classification could be for example a solution inspired by anomaly detection where the anomaly could be anything but noise (where the noise simply masks a non-variable star and everything else would be considered variable).

References

1. Andrešič, D., Šaloun, P., Suchánová, B.: Large astronomical time series preprocessing and visualization for classification using artificial neural networks. In: 2019 IEEE 15th International Scientific Conference on Informatics, pp. 000311–000316 (2019)
2. Armstrong, D.J., et al.: K2 Variable Catalogue I: A Catalogue of Variable Stars from K2 Field 0. 2014. arXiv: 1411.6830 [astro-ph.SR]
3. Armstrong, D.J., et al.: K2 variable catalogue i: A catalogue of variable stars from k2 field 0. In: Astronomy & Astrophysics 579, June 2015. A19. ISSN 1432-0746. https://doi.org/10.1051/0004-6361/201525889
4. Bagnall, A., et al.: Time-series classification with cote: The collective of transformation-based ensembles. IEEE Trans. Knowl. Data Eng. **27**(9), 2522–2535 (2015). https://doi.org/10.1109/TKDE.2015.2416723. ISSN 2326-3865
5. Bagnall, A., et al.: The great time series classification bake off: a review and experimental evaluation of recent algorithmic advances. Data Min. Knowl. Disc. **31**(3), 606–660 (2016). https://doi.org/10.1007/s10618-016-0483-9
6. Bahdanau, D., Cho, K., Bengio, Y.: Neural machine translation by jointly learning to align and translate. arXiv: 1409.0473 [cs.CL] (2014)
7. Bass, G., Borne, K.: Supervised ensemble classification of Kepler variable stars. In: Monthly Notices of the Royal Astronomical Society, vol. 459, April 2016, stw810. https://doi.org/10.1093/mnras/stw810
8. Batista, G.E.A.P.A., et al.: CID: an efficient complexity-invariant distance for time series. Data Min. Knowl. Disc. **28**(3), 634–669 (2013). https://doi.org/10.1007/s10618-013-0312-3
9. Box, G.E.P., Jenkins, G.M., Reinsel, G.C.: Time Series Analysis. Wiley (2008). https://doi.org/10.1002/9781118619193
10. van Cleve, J.E., et al.: Kepler: a search for terrestrial planets - Kepler data characterization handbook (2016)
11. Cui, Z., Chen, W., Chen, Y.: Multi-scale convolutional neural networks for time series classification. arXiv: 1603.06995 [cs.CV] (2016)
12. Anh Dau, H., et al.: The UCR Time Series Archive. arXiv: 1810.07758 [cs.LG] (2018)
13. Elman, J.L.: Finding structure in time. Cogn. Sci. **14**(2), 179–211 (1990). https://doi.org/10.1207/s15516709cog1402_1
14. Fawaz, H.I., Forestier, G., Weber, J., Idoumghar, L., Muller, P.A.: Deep learning for time series classification: a review. Data Min. Knowl. Disc. **33**(4), 917–963 (2019). https://doi.org/10.1007/s10618-019-00619-1
15. Fukushima, K.: Neocognitron: a self-organizing neural network model for a mechanism of pattern recognition unaffected by shift in position. Biol. Cybern. **36**(4), 193–202 (1980). https://doi.org/10.1007/bf00344251
16. Gamboa, J.C.B.: Deep learning for time-series analysis. arXiv: 1701.01887 [cs.LG] (2017)
17. Giles, C.L., Lawrence, S., Tsoi, A.C.: Mach. Learn. **44**(1/2), 161–183 (2001). https://doi.org/10.1023/a:1010884214864
18. He, K., et al.: Deep residual learning for image recognition. arXiv: 1512.03385 [cs.CV] (2015)
19. Hinners, T.A., Tat, K., Thorp, R.: Machine learning techniques for stellar light curve classification. Astron. J. **156**(1), 7 (2018). https://doi.org/10.3847/1538-3881/aac16d. ISSN 1538-3881

20. Hochreiter, S.: Untersuchungen zu dynamischen neuronalen Netzen. Diploma thesis, Institut für Informatik, Lehrstuhl Prof. Brauer, Technische Universität München (1991)
21. Hochreiter, S., Schmidhuber, J.: Long short-term memory. Neural Comput. **9**(8), 1735–1780 (1997). https://doi.org/10.1162/neco.1997.9.8.1735
22. Hosenie, Z., et al.: Comparing multiclass, binary, and hierarchical machine learning classification schemes for variable stars. Mon. Not. R. Astron. Soc. **488**(4), 4858–4872 (2019). https://doi.org/10.1093/mnras/stz1999. ISSN 1365-2966
23. Hu, B., Chen, Y., Keogh, E.J.: Time series classification under more realistic assumptions. In: SDM (2013)
24. Ioffe, S., Szegedy, C.: Batch normalization: accelerating deep network training by reducing internal covariate shift. arXiv: 1502.03167 [cs.LG] (2015)
25. Jenkins, J.M.: Kepler Data Processing Handbook: Overview of the Science Operations Center. Kepler Science Document, January 2017
26. Karim, F., Majumdar, S., Darabi, H., Chen, S.: LSTM fully convolutional networks for time series classification. IEEE Access **6**, 1662–1669 (2018). https://doi.org/10.1109/ACCESS.2017.2779939
27. Kennedy, J., Eberhart, R.: Particle swarm optimization. In: Proceedings of ICNN 1995 - International Conference on Neural Networks, vol. 4, pp. 1942–1948, November 1995. https://doi.org/10.1109/ICNN.1995.488968.
28. Lecun, Y., Bengio, Y.: Convolutional networks for images, speech, and time-series, January 1995
29. LeCun, Y., et al.: Handwritten digit recognition with a back-propagation network. In: Touretzky, D.S. (ed.) Advances in Neural Information Processing Systems 2, pp. 396–404. Morgan-Kaufmann (1990). http://papers.nips.cc/paper/293-handwritten-digit-recognition-with-a-back-propagation-network.pdf
30. LeCun, Y., et al.: Object recognition with gradient-based learning. In: Shape, Contour and Grouping in Computer Vision, pp. 319–345. Springer, Heidelberg (1999). https://doi.org/10.1007/3-540-46805-6_19
31. van der Maaten, L., Hinton, G.: Visualizing data using t-SNE. J. Mach. Learn. Res. **9**, 2579–2605 (2008). http://www.jmlr.org/papers/v9/vandermaaten08a.html
32. Petitjean, F., et al.: Dynamic time warping averaging of time series allows faster and more accurate classification. In: 2014 IEEE International Conference on Data Mining, pp. 470–479, December 2014
33. Rumelhart, D.E.: Chapter parallel distributed processing, exploration in the microstructure of cognition (1986)
34. Russakovsky, O., et al.: Imagenet large scale visual recognition challenge. arXiv: 1409.0575 [cs.CV] (2014)
35. Sakoe, H., Chiba, S.: Dynamic programming algorithm optimization for spoken word recognition. IEEE Trans. Acoust. Speech Signal Process. **26**(1), 43–49 (1978). https://doi.org/10.1109/TASSP.1978.1163055. ISSN 0096-3518
36. Sammon, J.: A nonlinear mapping for data structure analysis. IEEE Trans. Comput. **C–18**(5), 401–409 (1969). https://doi.org/10.1109/t-c.1969.222678
37. Samus', N.N., et al.: General catalogue of variable stars: Version GCVS 5.1. Astron. Rep. **61**(1), 80–88 (2017). https://doi.org/10.1134/s1063772917010085
38. Schäfer, P.: The BOSS is concerned with time series classification in the presence of noise. Data Min. Knowl. Disc. **29**(6), 1505–1530 (2014). https://doi.org/10.1007/s10618-014-0377-7
39. Schreiber, T., Schmitz, A.: Discrimination power of measures for nonlinearity in a time series. Phys. Rev. E **55**(5), 5443–5447 (1997). https://doi.org/10.1103/physreve.55.5443

40. Skoda, P.: Optical spectroscopy with the technology of virtual observatory. Baltic Astronomy 20 (2011). https://doi.org/10.1515/astro-2017-0332
41. Smirnov, D., Nguifo, E.M.: Time series classification with recurrent neural networks (2018)
42. Sutskever, I., Vinyals, O., Le, Q.: Sequence to sequence learning with neural networks. In: Advances in Neural Information Processing Systems, vol. 4, September 2014
43. Talbi, E.G.: Optimization of deep neural networks: a survey and unified taxonomy. Working paper or preprint, June 2020. https://hal.inria.fr/hal-02570804
44. Vanderburg, A.: K2 extracted lightcurves ("k2sff") (2015). http://archive.stsci.edu/doi/resolve/resolve.html?doi=10.17909/T9BC75
45. Wang, Z., Yan, W., Oates, T.: Time series classification from scratch with deep neural networks: a strong baseline. arXiv: 1611.06455 [cs.LG] (2016)
46. Williams, R.J., Zipser, D.: A learning algorithm for continually running fully recurrent neural networks. Neural Comput. **1**(2), 270–280 (1989). https://doi.org/10.1162/neco.1989.1.2.270

Data and Information Modelling

Requirements for Information Modelling in Manufacturing

Václav Jirkovský[1(✉)], Marek Obitko[2], Ondřej Šebek[1], and Petr Kadera[1]

[1] Czech Institute of Robotics, Informatics, and Cybernetics, Czech Technical University in Prague, Jugoslávských partyzánů 1580/3, 160 00 Prague, Czech Republic
{vaclav.jirkovsky,ondrej.sebek,petr.kadera}@cvut.cz

[2] Rockwell Automation R&D Center, Argentinská 1610/4, 170 00 Prague, Czech Republic
mobitko@ra.rockwell.com

Abstract. The world of industrial automation went through digital transformation in the last decades. This process brought myriads of software tools and systems that are usually incompatible, frequently due to the usage of different information and data models. The globalized market and development of e-commerce brought the trend of shorter product life-cycle, the need for prompt reactions to market needs, higher degree of product customization and higher degree of production automation. The industry responded with a new production paradigm: Industry 4.0. One of the key concepts of Industry 4.0 is the seamless integration of various information resources providing a ubiquitous knowledge-management system. The information integration is tightly connected to the interoperability of various applications and systems. The interoperability is usually achieved by information exchange. In this paper, the requirements for information models from the application perspective are presented and discussed. Next, the overview of possible formats for information exchange and standards and approaches for information modelling is provided. The formats for information exchange and way of information modelling are crucial for the resulting capabilities of the overall environment. The aim of this paper is to discuss and illustrate information exchange and modelling and to emphasize the importance and implications of this challenge.

Keywords: Industrial automation · Information exchange · Information modelling · Semantics

1 Introduction

The industrial automation domain has been influenced by the massive adoption of information and communication technologies in recent years, and this trend is represented by the 4th industrial revolution (Industry 4.0) [15]. Industry 4.0

J. Paralič et al. (Eds.): DISA 2020, AISC 1281, pp. 147–159, 2021.
https://doi.org/10.1007/978-3-030-63872-6_7

vision depends on digitization and virtualization of all automation levels in the factory from the field level to the enterprise level. Massive digitization and virtualization (including, for example, the exploitation of virtual twins) require connecting information from various areas as well as domains, e.g., information integration and exchange from overall product life-cycle, exploitation of simulations, integration of information about production processes, hardware capabilities and properties, as well as information from the supply chain.

The integration of such heterogeneous environments is essential for enabling the vision of Industry 4.0. In a nutshell, the integration resides in proper interpretation and understanding of information models within a given context as well as securing information exchange to enable faultless (from a conceptual point of view) communication. This task may be very challenging because of semantic and semiotic heterogeneity [13] and depends directly on formalism used for expressing information models.

Obviously, if the information model is not explicitly specified and is "encoded" within control algorithm of application, then the integration and proper understanding of information produced by the application is complicated. This situation is common when attempting to integrate usual legacy systems which store only data without any explicit specification of data meaning.

Furthermore, not only data definition but also the data related interpretation is often hard-coded and distributed across overall control algorithm, and thus any maintenance and change management is very difficult and sometimes close to impossible. On the other hand, explicitly specified information models or at least description of data (e.g., in the form of metadata) improve possible systems integration or interoperability. An example of this effort is the Model-Driven Architecture introduced in the last decade of the last century.

The information models and their interactions with surrounding systems are tightly connected to information exchange and formats which are exploited for the exchange. The formats differ by the expressivity they provide and by capabilities of how the format is "self-describing".

This paper is organized as follows: first, the requirements for information models are described. Next, the overview of the information exchange and information modeling is introduced and discussed. Finally, the applicability of different formats for creating the information model is demonstrated on simple use-case.

2 Requirements on Information Modelling and Information Exchange

Industry 4.0 paradigm, as briefly described above, assumes increasing digitization of all automation levels within a factory. One of the crucial and the most challenging goal is to allow different parts and levels of the system (machines, applications) to communicate and perform actions without the supervision of a human. To achieve this, particular components must be able to

- easily exchange information,
- interpret and analyze provided information, and
- retrieve information required for them to perform actions.

The basis for the tasks mentioned above is provided by the information model of the system. An information model may be defined as a specification of a given conceptualization, i.e., represents relevant concepts, their relations, and attributes. At the same time, the information model is designed by a human, and it also serves as a format for providing information to a human. Therefore, one of the characteristics of an information model could be its **understandability** by a user. Moreover, the model should be **easy to modify** or extend to facilitate information models reusability. Furthermore, models should also reflect the latest trends in manufacturing of Industry 4.0 such as flexible manufacturing and provide means for product-driven manufacturing [19]. All these features thus form requirements on the information model, its format and design as well as a set of supportive tools.

Understandability of Information Models — The ability of the systems to interpret or understand the information model depends mainly on the language/format used for the creation of the model. Particular languages/formats differ in their provided semantic expressivity as well as supportive tools, for example, means for querying, etc. The sufficient expressivity depends on application needs, e.g., ability to express restrictions, cardinality, ranges, etc.

Change Management of Information Models — The requirement for easy change management determines mainly the design of the information model. Regarding this feature, the very advantageous approach is to design the information model in an object-oriented way. In such an approach, the information is divided into smaller blocks or objects. Complex systems can be then described by relating and nesting lower-level information objects. This aspect is very close to modularity of information model.

Consistency Management of Information Models — To facilitate change management of the information model, it is necessary to ensure consistency of the modifications with respect to both data/file format as well as a logical format of the information model. The former task may be provided by schema validation tools. These tools are used to ensure syntactical and terminological homogeneity. The latter task (consistency maintenance while making modifications, regarding the logical format of the information model) may be provided by reasoning tools. Reasoning checks that the modifications do not violate any axiom of the information model.

Reasoning tools may also be used to infer new (hidden) knowledge based on existing knowledge. This gives the user (both human and machine/application) the capability of analysing the model and using implicit knowledge (knowledge not explicitly stated in the model by the designer) for further tasks.

Querying of Information Models — In the case of information models, where information is not clearly specified in not easy user-readable form even distributed across several locations, it is difficult to acquire required information from the system. In other words, it is not straightforward to retrieve a specific piece of information needed by a human or the machine to perform some action. For this reason, it is beneficial if the information model may be queried using some querying language, and if a tool or software library implementation is available to perform such querying.

Identification of Elements Within Information Models — As already mentioned, information sharing is important for the integration of various information models to form the interoperable system. The obstacle for faultless integration may be non-unique identification of elements of models. In some models, identification is related to a given local model, and in languages such as OWL, identification is unique because of the exploitation of URIs[1]. In some formats, GUIDs[2] can be recommended, but may not be always required.

Thus, the requirements on information models may be summarized (but not limited) as follows:

- Understandability of Information Models,
- Change Management of Information Models,
- Consistency Management of Information Models,
- Querying of Information Models,
- Identification of Elements within Information Models.

3 Information Models and Formats for Information Exchange

Information models are essential for every application. They maintain the most valuable things — information as well as their relations and limitations. In general, two types of information models may be distinguished

- **Non-explicit ("hard-coded") information models** are common to be used in prevalent part of applications. The non-explicit information model is represented by a logical structure of a control algorithm which utilizes relevant variables to achieve demanded objectives of the algorithm. If all of the required information and data for an application are implicitly hard-coded within a control code of an application, then it may be very efficient regarding time complexity of the control algorithm, but it is unsuitable regarding interoperability of applications.
- **Explicitly specified information models** represent explicit specification of given conceptualizations [11]. The explicitly specified information models stand aside of given control algorithm and required information for algorithm

[1] URI—https://cs.wikipedia.org/wiki/Uniform_Resource_Identifier.
[2] GUID—Globally Unique Identifier.

operation may be obtained with the help of querying the given information model.

Except for a type of selected formalism for information model, the application is also influenced by exploited format for information exchange because it limits possible interactions with a given neighbourhood. In other words, every application operates with respect to a given conceptualization, which is specified within a related information model, and application neighbourhood which represents a given context (in many cases). Thus, information exchange is strongly related with information models of applications, i.e., a format for the exchange has to be able to ensure communication between all of the models.

In the following section, information modelling formats for information exchange are described.

3.1 Information Modeling

Information models represent backbones of every system, and their formalism influences their capabilities as well as behaviour. An information model may be incorporated directly into the application algorithm or maybe explicitly specified and stand separated from the application algorithm. Separated or "hard-coded" information models may be considered as a specification of a conceptualization relevant to a given domain or context.

In the past, information models were not specified explicitly within the applications coded in procedural programming languages. The model of such an application was directly hard-coded in a given algorithm. This approach is often effective; however, every change, reusability, or even sharing of this type of information model was difficult and sometimes close to impossible.

The process of design of explicitly specified information models is shown in Fig. 1. There are many various approaches how to design information models. According to [12], the process of information modelling consists of 6 steps — scenario identification, informal competency questions (denoting problem which should be solved). Next steps are creation of terminology and formal competency questions specification. The last step is the definition of axioms.

The development of the information model is motivated by a specific scenario. Prior to the information model design it is also necessary to define problems or questions that the information model should be able to solve or answer (informal competency questions step). The next step comprises definition of the terminology, i.e., definition of objects, attributes and relations needed to describe the problem or questions within the information model. Next step defines the formal competency questions, i.e., translates the informal competency questions using the terminology defined in the previous step. This step determines which axioms should be included in the ontology. Finally the axioms are defined in the information model. The axioms provide meaning to the terms (objects) specified in the information model, i.e., define relations to other objects, constraints on the object, etc.

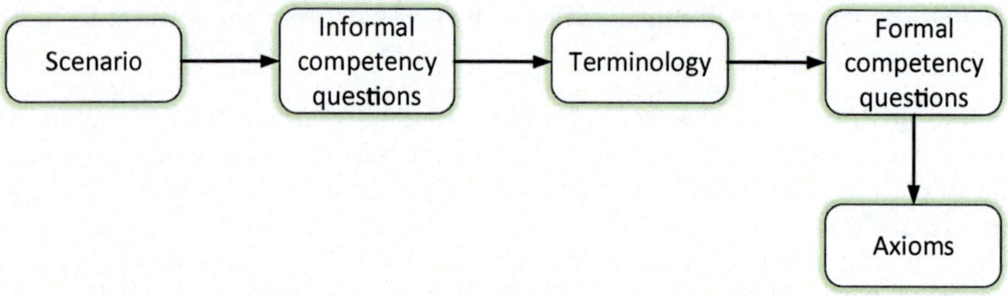

Fig. 1. Process of information modelling

In come cases, there may already exist an information model describing part of the target domain. In such case the recommended approach is to build the new information model upon the existing one. This approach facilitates reusability and shareability of the new model for other applications or users that may already have experience with the existing model.

3.2 Formats for Information Exchange

There are many various formats for information exchange applicable to various application domains. Originally plain text or proprietary binary files were used for information exchange between applications. Over time, more elaborated methods and formats appeared including CSV [24], XML [4], JSON [7], JSON-LD [25], RDF [18], and more. Choice of particular format depends on intended application. Different formats may be considered when exchanging pure data or parts of the information model itself, in some applications speed of the transfer may be critical aspect so that verbosity of given format must be taken into account.

Various formats may be categorized according to many criteria. The most fundamental criteria is the structure of the format. On one side there are unstructured formats such as plain text or binary files. There are no rules how such file should be composed. Structured formats define rules on valid composition or structure of the file. Structure of the file is determined using some reserved characters, keywords, tags, or indentation. One of the simpler structured formats is for example CSV (Comma Separated Values) which only prescribes the values of different variables on single line to be separated by commas. The formats such as plain text or CSV are rather suitable for exchange of pure data (e.g., time series of recordings).

There are other advanced structured formats which allow to define more complex data structure or objects. Data in these formats are split into elements which may be nested into each other. Some examples of commonly used formats are XML (eXtensible Markup Language)[3], JSON (JavaScript Object Notation), or YAML (YAML Ain't Markup Language). The advantage is that these formats

[3] https://www.w3.org/standards/xml.

are relatively easy to process by the machine and understandable by humans. Due to the object-like nature they can be also easily modified or extended.

Some formats, including XML, JSON, are supplemented with so called schemas, which define valid structure and content of given document, including for example definition of allowed datatypes, nesting of elements, their order, etc. This feature is very beneficial regarding the change management and consistency management requirements on the information model.

Based on the generic formats for information exchange, many various company-specific formats have been designed and developed. However, these formats are suitable for information exchange only with respect to a limited set of applications or systems and their expressivity is limited by this set of applications, and their information models. Their utilization in a general heterogeneous environment is relatively hard or impossible.

One example of such format from the manufacturing domain is PLM XML created by Siemens PLM to facilitate an exchange of product lifecycle data. Another example of the industry specific format is the AutomationML [6]. It is the emerging IEC 62714-based standard intended for facilitation of uniform data exchange between engineering tools. It uses XML for serialization. AutomationML is a data exchange format which incorporates different standards such as:

- CAEX [9] for topology — hierarchical structure of properties and relations of objects.
- COLLADA [1] for geometry (graphical attributes and 3D information) and kinematics.
- PLCopen XML [8] for the control application of the plant.

3.3 Formats for Information Modelling

Besides the formats for information/data exchange there are standards designed specifically for the purpose of information modelling. Within the context of manufacturing domain the most common are OPC UA standard and ontologies.

OPC Unified Architecture (OPC UA) standard [20]. In general, OPC UA is a secure and open mechanism for exchanging data between servers and clients in industrial automation as well as for information modelling. OPC UA standard tries to overcome the main obstacle of its predecessor (OPC Data Access together with OPC Historical Data Access and OPC Alarm&Events) — COM[4] dependency of OPC. Therefore, the OPC UA was designed for the replacement of all existing COM-based specification to be platform-independent with extensible modelling capabilities.

OPC UA is built on two main components [22] — transport mechanisms and data modelling. The transport component offers the possibility to communicate via optimized binary TCP protocol for high-performance intranet communication and the next possibility to communicate via Web Services. The data

[4] https://www.microsoft.com/com/default.mspx.

modelling component represents rules and building blocks for the creation and exposing information model. It also defines, for example, base types for building a type hierarchy.

In other words, OPC UA information modelling framework has completed object-oriented capabilities. The information model itself is unambiguously described, and the meaning of information is machine-understandable with the help of shared contextual information, i.e., objects have specified properties as well as relations to other objects. Unfortunately, OPC UA currently lacks a more strict self-describing explicit definition of its entities, i.e., explicitly define assertions or axioms on entities and references in addition to the implicitly defined their mathematical properties [2]. Furthermore, currently it lacks means for the unique identification across more information models, and therefore the entities may be used confusingly.

Furthermore, OPC UA does not provide any universal syntax for modelling an information model. In general, every SDK or stack implementation offers a different way on how to model the information model. The information to exchange is uniformed by OPC UA Stack to work across different OPC UA stack implementation.

Ontologies are exploited for description and serialization of a given conceptualization (a mental image of some domain). Specifically, the term ontology may be defined using the well-known definition from Thomas Gruber [11] — an ontology is "*explicit specification of a shared conceptualization.*"

The cornerstones of semantic web technologies are Resource Description Framework (RDF), Resource Description Framework Schema (RDFS), and Web Ontology Language (OWL). RDF is a framework for describing resources, where resources may be anything, e.g., various physical or abstract concepts. Resources are described using statements with the structure of a triple — (subject, predicate, object). RDFS provides a semantic extension of RDF — mechanisms for describing groups of related resources and the relationships between these resources. RDFS is written in RDF using an extended set of keywords. In other words, RDF defines mainly a general syntax of statements (triples), and RDFS adds a semantics. Web Ontology Language (OWL) is a Semantic Web standard intended to represent complex knowledge about resources. OWL enables richer semantic models by defining additional keywords to express complex features of ontology elements including for example relations between classes and individuals (e.g., disjoint classes, equal individuals), attributes of properties (e.g., transitive, symmetric properties) and their cardinality (e.g., minimum, maximum, exact) or class property restrictions (e.g., all values from).

OWL is a description logic based language designed so that it is possible to do a reasoning by a computer programs either to verify the consistency of that knowledge or to make implicit knowledge explicit.

Ontologies have been proven to be suitable tool for facilitating various tasks within the manufacturing domain, including for example product lifecycle managemenet [5], service discovery [14], analysing of change influences in manufacturing systems [10], etc. Various ontologies for the manufacturing domain have been proposed, e.g., Manufacturing Core Concept Ontology [27], Manufacturing's Semantics Ontology [21], Adaptive holonic control architecture for distributed manufacturing system ontology [3] or AutomationML Ontology [17].

4 Demonstration

In this section, we discuss the applicability of the OPC UA and OWL formats and standards for creating information models. For this purpose, we assume a simple use case of designing an information model for the description of industrial robots. Figure 2 displays example of such information model implemented using OPC UA.

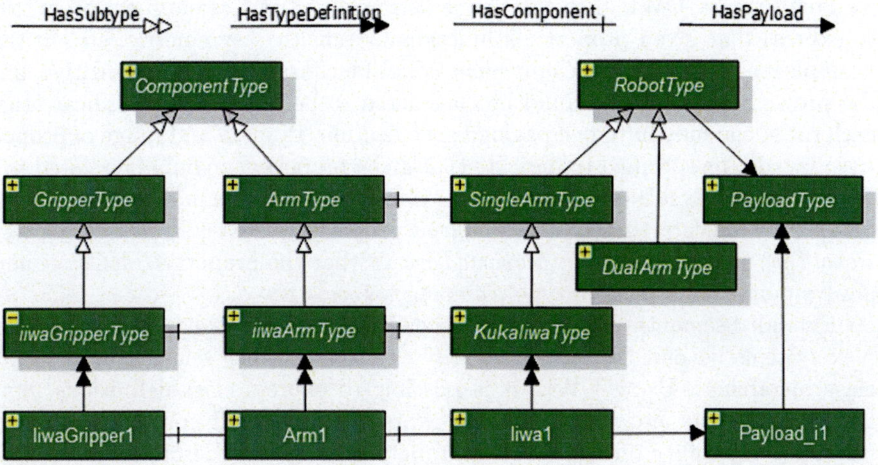

Fig. 2. Graphical representation of sample robot information model expressed in OPC UA standard.

This model defines basic class hierarchy and relations between classes, instances of classes and relations between the instances. In OPC UA classes are represented using the *ObjectType* nodes and individuals are represented using the *Object* nodes. The relation *HasSubtype* is used to define class hierarchy and the relation *HasTypeDefinition* to assign individual to given class. The relation *HasComponent* is used to describe the structure of the robot. These three relations are native OPC UA references defined in the *ReferenceType* node class. In this example, the model describes some characteristics of the robot (payload) and relates these characteristics to the robot class/individual using the *HasPayload* relation.

OPC UA provides rather basic semantics suitable mainly to define class hierarchies and is-a relationships as displayed in Fig. 2. OWL language implements much richer semantics and allows to define, for example, class restrictions, relations between classes and individuals, the cardinality of properties and their attributes, domain and range. Using these constructs, the simple robot information model from Fig. 2 can be significantly enriched.

For instance, class restrictions may be used to define the meaning of a particular class, i.e., specify the conditions given object must fulfil to be considered as an instance of the class. For example, the classes of single-arm and dual-arm robots could be defined as subclasses of robots that have one/two arms using the cardinality restriction (e.g., HasComponent exactly 2 ArmType for dual-arm robot class). Similarly, the restriction could be put, for example, on the *iiwaGripperType* class to define what the characteristics of gripper used for iiwa robot are. Moreover, OWL allows describing other relations between classes. For example, the two subclasses of the *RobotType* class or two subclasses of the *ComponentType* could be labeled as disjoint classes.

Furthermore, OWL allows to define extra characteristics of object properties to express that given property is functional, transitive, symmetric, etc. In this example, the property *HasComponent* could be considered to be transitive, and the property *HasPayload* could be considered to be functional to indicate that each robot can only have one payload specification. Domain and range of properties may also be specified to indicate that given property can only be related to a particular object/subject class. In this example, the domain and the range could be used to indicate that the *HasPayload* property relates only objects of type *RobotType* to objects of type *PayloadType*, or that the property *HasComponent* always points on objects of type *ComponentType*.

Extended semantic expressivity of the OWL language allows describing complex systems beyond the level of specifying class/instance relations and simple class hierarchies. Using OWL, it is possible to express class definitions, limitations of objects, etc. This extended expressivity naturally provides extended reasoning capability on the OWL information model. As described above, the reasoning is important way facilitating consistency management of the information model as it helps to check that no axioms are violated when modifying or extending the information model.

Furthermore, the advantage of OWL is that various supporting tools are available. For example, different reasoners with different capabilities may be used on the OWL model. On the other hand, there are no reasoners available for the OPC UA model. This limitation can be formally circumvented by transforming the OPC UA model into ontology description and applying the ontology reasoner as demonstrated in [2]. Furthermore, there are several query languages for OWL models, e.g., SPARQL or SQWRL. The query languages may be used to retrieve information from the model and perform tasks such as semantic matchmaking, i.e., to search objects that satisfy given conditions. This can be used, for example, to find resources (e.g., grippers/robots) suitable to perform operations based on prescribed specifications as demonstrated in [14].

From this simple example, it is obvious that OWL is semantically richer than OPC UA. The lack of the better expressivity of OPC UA is described in [23].

However, the richer expressivity of OWL has several drawbacks. The most significant drawback is the performance issue because of processing more information compared to OPC UA. Furthermore, a possible obstacle for the massive exploitation of OWL together with reasoning within the industrial automation domain may be caused by Open Word Assumption [26], which could be difficult to handle by many developers and engineers.

5 Discussion and Conclusions

In this paper, some of the essential requirements of applications and systems on the information modelling regarding Industry 4.0 paradigm are presented. The fulfilment of the requirements heavily depends on the utilized formalism of a given information model as well as on communication format for information exchange. Thus, the overview of the requirements is followed by a brief overview of the selected significant formats for information exchange, together with standards and approaches for information modelling.

The formats for information exchange affect how expressively information is shared between the systems. The format has to be accepted across the overall ecosystem of integrated and interoperable systems and frameworks. Many of available formats may exploit a related schema which is able to apply a set of rules to which a given document must conform in order to be considered "valid" according to that schema.

A selected formalism and approach for the information modelling affect many characteristics of a given system, e.g., interoperability capabilities of the system, its performance, means for reasoning in the model, querying the model, change management capabilities, etc. All of these characteristics represent requirements on the system and have to be assessed according to a given deployment. Furthermore, these requirements have to be taken into consideration during the design phase of the system architecture by a developer or a system engineer.

In the overview of standards for information modelling, a relational database schema [16] is missing because there are a few systems which are schema driven. However, it is worth noticing the relational database schema because it is important for information sharing, i.e., various applications may utilize the same relational database and thus enable or facilitate information sharing. On the other hand, the meaning of information may be different across these applications, and it could be the cause of semantic heterogeneity.

In this relatively brief overview of available formats and standards, it is not possible to cover all of their specific properties and characteristics. For example, OPC UA information model provides a way how to explicitly specify conceptualization and use the model within the application for its control. However, this specification is (in many of available OPC UA stack and SDK implementations) distributed across the overall implementation of the OPC UA server, and thus sharing or reusability of the model is relatively difficult. The information model or its address space may be easily shared only by OPC UA communication between OPC UA clients and servers.

The aim of this paper is not to prove or show that one of the formats or standards is the best and should be used in every system and application. The aim is to give a short introduction of requirements of systems on information modelling and exchange formats to a reader. The knowledge engineer or developer has to find out the trade-off between rich expressiveness of the given system and system performance, and this selection every time depends on a given context. Furthermore, the skills of users and operators have to be also taken into consideration. Thus, if the developer demands rich expressiveness of the information model formalism and easy system interoperability together with its reusability, then standards such as OWL should be chosen for the information model. On the other hand, if the developer requires high-performance of the system regardless, for example, to its bad reusability and change management, then the developer will choose or develop some proprietary format or architecture.

Acknowledgment. This research has been supported by Rockwell Automation Laboratory for Distributed Intelligent Control (RA-DIC), by institutional resources for research by the Czech Technical University in Prague, Czech Republic, and by the OP VVV DMS Cluster 4.0 project funded by The Ministry of Education, Youth and Sports.

References

1. Arnaud, R., Barnes, M.C.: COLLADA: Sailing The Gulf of 3D Digital Content Creation. CRC Press, Boca Raton (2006)
2. Bakakeu, J., Brossog, M., Zeitler, J., Franke, J., Tolksdorf, S., Klos, H., Peschke, J.: Automated reasoning and knowledge inference on OPC UA information models. In: 2019 IEEE International Conference on Industrial Cyber Physical Systems (ICPS), pp. 53–60. IEEE (2019)
3. Borgo, S., Leitão, P.: Foundations for a core ontology of manufacturing. In: Sharman, R., Kishore, R., Ramesh, R. (eds.) Ontologies: A Handbook of Principles, Concepts and Applications in Information Systems, pp. 751–775. Springer, Boston (2007). https://doi.org/10.1007/978-0-387-37022-4_27
4. Bray, T., Paoli, J., Sperberg-McQueen, C.M., Maler, E., Yergeau, F., et al.: Extensible markup language (xml) **1** (2000)
5. Bruno, G., Antonelli, D., Villa, A.: A reference ontology to support product lifecycle management. In: 9th CIRP Conference on Intelligent Computation in Manufacturing Engineering - CIRP ICME 2014 Procedia CIRP, vol. 33, pp. 41–46 (2015). http://www.sciencedirect.com/science/article/pii/S2212827115006526
6. Commission, I.E., et al.: Engineering data exchange format for use in industrial automation systems engineering-automation markup language-part 1: architecture and general requirements. Int. Std. Rev. **1**, 1–62714 (2014)
7. Crockford, D.: Introducing json. http://www.json.org (2009)
8. Estévez, E., Marcos, M., Lüder, A., Hundt, L.: Plcopen for achieving interoperability between development phases. In: 2010 IEEE 15th Conference on Emerging Technologies & Factory Automation (ETFA 2010), pp. 1–8. IEEE (2010)
9. Fedai, M., Drath, R.: Caex-a neutral data exchange format for engineering data. ATP Int. Autom. Technol. **1**(2005), 3 (2005)

10. Feldmann, S., Kernschmidt, K., Vogel-Heuser, B.: Combining a sysml-based modeling approach and semantic technologies for analyzing change influences in manufacturing plant models. Procedia CIRP 17, 451 – 456 (2014), http://www.sciencedirect.com/science/article/pii/S2212827114004181, variety Management in Manufacturing
11. Gruber, T.R.: Toward principles for the design of ontologies used for knowledge sharing? Int. J. Hum. Comput. Stud. **43**(5), 907–928 (1995). http://www.sciencedirect.com/science/article/pii/S1071581985710816
12. Grüninger, M., Fox, M.S.: Methodology for the design and evaluation of ontologies (1995)
13. Jirkovský, V.: Semantic integration in the context of cyber-physical systems (2017)
14. Jirkovský, V., Šebek, O., Kadera, P., Burget, P., Knoch, S., Becker, T.: Facilitation of domain-specific data models design using semantic web technologies for manufacturing. In: Proceedings of the 21st International Conference on Information Integration and Web-Based Applications & Services, pp. 649-653 (2019)
15. Kagermann, H., Wahlster, W., Helbig, J.: Securing the future of german manufacturing industry. Recommendations for Implementing the Strategic Initiative INDUSTRIE **4**(199), 14 (2013)
16. Kanellakis, P.C.: Elements of relational database theory. In: Formal Models and Semantics, pp. 1073–1156. Elsevier (1990)
17. Kovalenko, O., Grangel-González, I., Sabou, M., Lüder, A., Biffl, S., Auer, S., Vidal, M.E.: Automationm ontology: modeling cyber-physical systems for industry 4.0. IOS Press Journal (2018)
18. Lassila, O., Swick, R.R., et al.: Resource description framework (RDF) model and syntax specification (1998)
19. Legat, C., Lamparter, S., Seitz, C.: Service-oriented product-driven manufacturing. IFAC Proc. Vol. **43**(4), 150–155 (2010)
20. Leitner, S.H., Mahnke, W.: OPC UA-service-oriented architecture for industrial applications. ABB Corp. Res. Cent. **48**, 61–66 (2006)
21. Lemaignan, S., Siadat, A., Dantan, J.., Semenenko, A.: Mason: A proposal for an ontology of manufacturing domain. In: IEEE Workshop on Distributed Intelligent Systems: Collective Intelligence and Its Applications (DIS 2006), pp. 195–200 (2006)
22. Mahnke, W., Leitner, S.H.: OPC Unified Architecture. Springer, Heidelberg (2009)
23. Pessemier, W.: Why semantics matter: or how to build smart machines using OPC UA (2015). https://opcconnect.opcfoundation.org/2015/12/why-semantics-matter/
24. Shafranovich, Y.: Common format and mime type for comma-separated values (csv) files (2005)
25. Sporny, M., Longley, D., Kellogg, G., Lanthaler, M., Lindström, N.: JSON-LD 1.0. W3C Recommendation **16**, p. 41 (2014)
26. Tao, J., Sirin, E., Bao, J., McGuinness, D.L.: Integrity constraints in owl. In: 24th AAAI Conference on Artificial Intelligence (2010)
27. Usman, Z., Young, R.I.M., Chungoora, N., Palmer, C., Case, K., Harding, J.: A manufacturing core concepts ontology for product lifecycle interoperability. In: van Sinderen, M., Johnson, P. (eds.) Enterprise Interoperability, pp. 5–18. Springer, Heidelberg (2011). https://doi.org/10.1007/978-3-642-19680-5_3

reasoning. Briefly, formal concept analysis scrutinizes an object-attribute block of relational data in bivalent form.

The mathematical models of machine learning and data analysis have been studied in terms of formal concept analysis. Kuznetsov [2] presented the relation of the model of machine learning from positive and negative examples (JSM-learning) to the well-known domains in data analysis and artificial intelligence: implications in formal concept analysis, search for functional dependencies in databases, decision trees, data mining and knowledge discovery.

Formal concept analysis is also an effective tool in data mining methods based on results from applied algebra, fuzzy logic, aggregation operators, or conceptual structures [3–5]. Actual research in the field of formal conceptual analysis is focused, for example, on reducing the number of clusters, factorizing matrices [6], or generating attribute implications [7]. One of the outputs of this method of data analysis are concepts (two-component clusters), which are formed by pairs of closed sets of objects and attributes [3]. We proposed several generalizations of these methods and applied them to different data sets [8–11].

In our research, we proposed a formal framework for research on formal concept analysis of higher order [9], which we applied in modeling highly heterogeneous tabular data of conference participants' preferences. We examined formal concept analysis using category theory in terms of relationships between fuzzy formal contexts. Our results point to the equivalence of categories of formal contexts and L-Chu correspondences and completely union L-ordered sets and isotonic Galois connections [12,13]. Our efforts led to categorical research in the field of the mentioned heterogeneous extension of formal concept analysis, which led to the results described in the articles [14] concerning *-autonomous categories and the related Chu construction. The mentioned concepts of category theory are important tools, for example, in the description of quantum systems or information flow in distributed systems [15].

The applications of formal concept analysis which focus on the text retrieval and text mining were published in the monograph [16]. The state of art of formal concept analysis and its applications in linguistics, lexical databases, rule mining (iceberg concept lattice), knowledge management, software engineering, object-engineering, software development and data analysis are covered in the special LNAI 3626 Volume devoted to Formal concept analysis – Foundations and Applications and edited by Berhard Ganter, Gerd Stumme and Rudolf Wille in 2005 [17]. An extensive overview of applications of formal concept analysis in the various application domains including software mining, web analytic, medicine, biology and chemistry data is given by [18]. Particularly, we mention the conceptual difficulties in the education of mathematics explored by [19], the techniques for analyzing and improving integrated care pathways [20], finding a set of representative symptoms for the disease [21] or evaluation of questionnaires presented in [22,23].

In [24], formal concept analysis is applied as a tool for image processing and detection of inaccuracies. Recently, the morphological image and signal processing from the viewpoint of L-fuzzy formal concept analysis are explicitly presented in [25]. The main results offer the possibility to interpret the binary images as the classical formal concepts and open digital signals as L-fuzzy formal concepts. Many other applications can be found in the literature, nevertheless there are still novel challenges for researchers.

The implications $Z \rightarrow W$ over set of attributes A, whereby Z and W are the subsets of A have been studied thoroughly in computer science and applied mathematics. In formal concept analysis, the implications $Z \rightarrow W$ over set of attributes A are called attribute implications [26]. In object-attribute tables, we can interpret these formulas by the statement that each object having all attributes from Z has also all attributes from W. From the Guigues-Duquenne basis for attribute implications of a formal context, we can generate all implications valid in a formal context. The most important property of this basis is that it has minimal size from all possible sets of attribute implications generating all attribute implications that are valid in a formal context [27–30].

The attribute implications have some connection to the problem of mining association rules which was introduced by Agrawal et al [31]. If we consider association rules in a formal context, then we can define the support as the number of objects which contain attributes from the premise. The confidence defines which percentage of objects that contain attributes from the premise of association rule also contain the attributes from the conclusion of the association rule. If confidence equals one, we can call it exact association rule in a formal context and it corresponds to the attribute implication valid in a formal context. Otherwise, we can call it partial attribute implication valid in a formal context or approximate association rule in a formal context [32–34].

The attribute implications and the process of classical attribute exploration are introduced in this chapter in order to their exploitation in our applications. The partial attribute implications which are connected to association rules are applied in this work for measuring of pupils' computational thinking.

This chapter is organized as follows. In Sect. 2, we present the computational methods for generation of attribute implications in the field of formal concept analysis. We illustrate the process of attribute exploration for special types of object-attributes tables in Sect. 3. The application of the partial attribute implications in formal concept analysis to analyze the real data from national testing of pupils regarding the computational thinking is thoroughly described in Sect. 4. We proposed the own test for measuring of pupils' computational thinking in several consecutive stages. The conclusions and several potential areas of further research conclude this chapter.

2 Attribute Exploration and Partial Attribute Implications

In this section, we define the structure of attribute implications basis [3,16,27] and describe the method of classical attribute exploration in formal concept analysis [28–30]. We introduce the necessary definitions and results which we in Sect. 3 and Sect. 4 devoted to our own examples and applications. We conclude this section by determining the attribute implications which can be generated from the concept lattices with reduced labeling.

Definition 1. *Let A be a non-empty set of attributes, $Z, W, Y \subseteq A$ and $\mathcal{Y} \subseteq \mathcal{P}(A)$. An attribute implication over A is an expression $Z \to W$. An attribute implication $Z \to W$ over A is valid in a set Y if $Z \subseteq Y$ implies $W \subseteq Y$ (or equivalently if $Z \nsubseteq Y$ or $W \subseteq Y$). An attribute implication $Z \to W$ over A is valid in $\mathcal{Y}$ if $Z \to W$ is valid in a set Y for every $Y \in \mathcal{Y}$.*

We will illustrate the definition on the set of attributes $A = \{\mathrm{i}, \mathrm{ii}, \mathrm{iii}\}$ throughout this section. The attribute implication $\{\mathrm{ii}\} \to \{\mathrm{i}\}$ over $A = \{\mathrm{i}, \mathrm{ii}, \mathrm{iii}\}$ is valid in $\mathcal{Y} = \{\emptyset, \{\mathrm{i}\}, \{\mathrm{iii}\}, \{\mathrm{i}, \mathrm{ii}\}, \{\mathrm{i}, \mathrm{iii}\}, \{\mathrm{i}, \mathrm{ii}, \mathrm{iii}\}\}$. In the propositional calculus, the attribute implication can be seen as the cumulative clause consisting of Horn clauses, which means the disjunction of attributes with at most one unnegated attribute.

The attribute implications can be fruitfully applied also in the view of formal concept analysis. It requires the definitions of central notions of a formal context and concept-forming operators.

Definition 2. *Let B and A be the nonempty sets and let $R \subseteq B \times A$ be a relation between B and A. A triple $\langle B, A, R\rangle$ is called a formal context, the elements of set B are called objects, the elements of set A are called attributes and the relation R is called incidence relation.*

The illustration of the one possible formal context for three objects, three attributes and incidence relation R is shown in Fig. 1.

	i	ii	iii
a	×		
b	×	×	
c			×

Fig. 1. A formal context $\langle\{\mathrm{a,b,c}\}, \{\mathrm{i,ii,iii}\}, R\rangle$

Definition 3. *Let $\langle B, A, R\rangle$ be a formal context and $X \in \mathcal{P}(B)$, $Y \in \mathcal{P}(A)$. Then the maps $\nearrow : \mathcal{P}(B) \to \mathcal{P}(A)$ and $\swarrow : \mathcal{P}(A) \to \mathcal{P}(B)$ defined by*

$$\nearrow(X) = X^{\nearrow} = \{y \in A : (\forall x \in X)\langle x, y\rangle \in R\}$$

and

$$\swarrow(Y) = Y^{\swarrow} = \{x \in B : (\forall y \in Y)\langle x, y\rangle \in R\}$$

are called concept-forming operators (also called derivation operators or polars) of a given formal context.

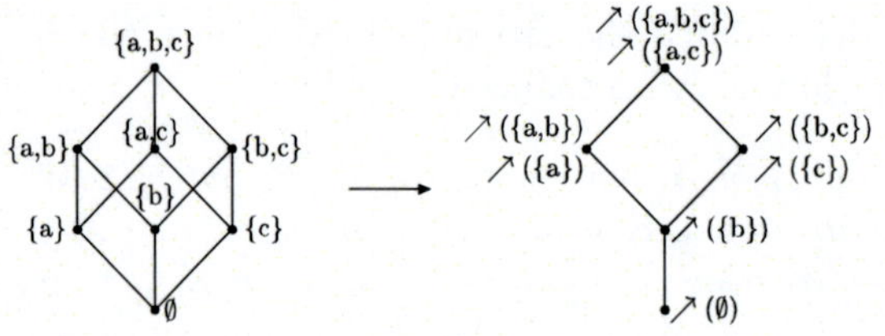

Fig. 2. Polar $\nearrow$ between $(\mathcal{P}(B), \subseteq)$ and $(\mathcal{P}(A), \supseteq)$

Figure 2 illustrates the concept-forming operators applied for a formal context described in Fig. 1.

Now, we can define the notion of attribute implications in the terms of formal context with the help of concept-forming operators.

Definition 4. *Let $\langle B, A, R\rangle$ be a formal context, $Z, W \subseteq A$ and $\mathcal{Y} \subseteq \mathcal{P}(A)$. An attribute implication $Z \to W$ over A is valid in $\langle B, A, R\rangle$ if $Z \to W$ is valid in $\mathcal{Y} = \{\{x\}^{\nearrow} : x \in B\}$.*

Consider a formal context in Fig. 1. Then the attribute implication $\{\text{ii}\} \to \{\text{i}\}$ over the set $A = \{\text{i}, \text{ii}, \text{iii}\}$ is valid in $\langle B, A, R\rangle$, because $\{\text{ii}\} \to \{\text{i}\}$ is valid in $\mathcal{Y} = \{\{\text{i}\}, \{\text{i}, \text{ii}\}, \{\text{iii}\}\}$. The attribute implication $\{\text{i}, \text{ii}\} \to \{\text{iii}\}$ over $A = \{\text{i}, \text{ii}, \text{iii}\}$ is not valid in $\langle B, A, R\rangle$, since $\{\text{i}, \text{ii}\} \to \{\text{iii}\}$ is not valid in $\{\text{i}, \text{ii}\}$ for $x = \text{b}$.

The natural requirement is to obtain the set of all attribute implications in a formal context. The notions of a model and a theory over the set of attributes are necessary for such consideration.

Definition 5. *Let T_A be an arbitrary set of attribute implications over the set A. Then T_A is called a theory over A. Let $Y, Z, W \subseteq A$. A set Y is called a model of theory T_A if every attribute implication from T_A is valid in Y. The set of all models of a theory T_A is denoted by* $\text{Mod}(T_A)$. *Let $Z \to W$ be the attribute implication and let $Z \to W$ be valid in every model Y from* $\text{Mod}(T_A)$. *Then we say that $Z \to W$ follows semantically from a theory T_A and denote by $T_A \models Z \to W$.*

Let $A = \{\text{i}, \text{ii}, \text{iii}\}$ and let a theory $T_A = \{\{\text{ii}\} \to \{\text{i}\}, \{\text{i}, \text{iii}\} \to \{\text{ii}\}\}$. The set of all models of this theory is $\text{Mod}(T_A) = \{\emptyset, \{\text{i}\}, \{\text{iii}\}, \{\text{i}, \text{ii}\}, \{\text{i}, \text{ii}, \text{iii}\}\}$. Consider the attribute implications $\{\text{ii}, \text{iii}\} \to \{\text{i}\}$ and $\{\text{i}\} \to \{\text{ii}\}$. It holds that $T_A \models \{\text{ii}, \text{iii}\} \to \{\text{i}\}$, but $T_A \not\models \{\text{i}\} \to \{\text{ii}\}$ because $\{\text{i}\} \to \{\text{ii}\}$ is not valid in $\{i\} \in \text{Mod}(T_A)$. To confirm that $T_A \models Z \to W$, we need to explore if $Z \to W$ is valid in every $Y \in \text{Mod}(T_A)$. We present a simplification of this process in the following.

Lemma 1. $\text{Mod}(T_A)$ *is a closure system on A.*

Proof. Since $Z, W \subseteq A$, then from a definition we have that every $Z \to W$ is valid in A. Therefore, all $Z \to W$ from T_A is valid in A and $A \in \text{Mod}(T_A)$. Let

$Y_i \in \mathrm{Mod}(T_A)$ for every $i \in I$. We want to prove that $\bigcap_{i\in I} Y_i \in \mathrm{Mod}(T_A)$. Let $Z \to W \in T_A$, then we want to prove that $Z \subseteq \bigcap_{i\in I} Y_i$ implies $W \subseteq \bigcap_{i\in I} Y_i$. From $Z \subseteq \bigcap_{i\in I} Y_i$ we have that $Z \subseteq Y_i$ for every $i \in I$, but this implies that $W \subseteq Y_i$ for every $i \in I$, since we suppose that $Y_i \in \mathrm{Mod}(T_A)$. Then it holds that $W \subseteq \bigcap_{i\in I} Y_i$ and we have $\bigcap_{i\in I} Y_i \in \mathrm{Mod}(T_A)$. □

Moreover, it can be proved that $\mathrm{cl}_{\mathrm{Mod}(T_A)}$ is a closure operator on A. It means that in order to check $T_A \models Z \to W$, it is sufficient to verify only if $Z \to W$ is valid in $\mathrm{cl}_{\mathrm{Mod}(T_A)}(Z) = \bigcap\{Y \in \mathrm{Mod}(T_A) : Z \subseteq Y\}$.

Consider $A = \{\mathrm{i}, \mathrm{ii}, \mathrm{iii}\}$ and theory $T_A = \{\{\mathrm{ii}\} \to \{\mathrm{i}\}, \{\mathrm{i}, \mathrm{iii}\} \to \{\mathrm{ii}\}\}$. Then $\mathrm{Mod}(T_A) = \{\emptyset, \{\mathrm{i}\}, \{\mathrm{iii}\}, \{\mathrm{i}, \mathrm{ii}\}, \{\mathrm{i}, \mathrm{ii}, \mathrm{iii}\}\}$. Let $\{\mathrm{ii}, \mathrm{iii}\} \to \{\mathrm{i}\}$ be the attribute implication. Then $\mathrm{cl}_{\mathrm{Mod}(T_A)}(\{\mathrm{ii}, \mathrm{iii}\}) = \{\mathrm{i}, \mathrm{ii}, \mathrm{iii}\}$ and since $\{\mathrm{ii}, \mathrm{iii}\} \to \{\mathrm{i}\}$ is valid in $\{\mathrm{i}, \mathrm{ii}, \mathrm{iii}\}$, we have $T_A \models \{\mathrm{ii}, \mathrm{iii}\} \to \{\mathrm{i}\}$. Moreover, one can use the elementary deduction rules and many other derivable rules from logic to find attribute implications which semantically follows from the given theory, but it requires the more detailed results about the soundness and completeness.

Definition 6. *Let T_A be the set of attribute implications and $Z \to W \in T_A$. Then T_A is called a non-redundant set of attribute implications if for every $Z \to W \in T_A$ holds $T_A \setminus \{Z \to W\} \not\models Z \to W$.*

The set $T_A = \{\{\mathrm{ii}\} \to \{\mathrm{i}\}, \{\mathrm{i}, \mathrm{iii}\} \to \{\mathrm{ii}\}, \{\mathrm{ii}, \mathrm{iii}\} \to \{\mathrm{i}\}\}$ is redundant, because $\{\{\mathrm{ii}\} \to \{\mathrm{i}\}, \{\mathrm{i}, \mathrm{iii}\} \to \{\mathrm{ii}\}\} \models \{\mathrm{ii}, \mathrm{iii}\} \to \{\mathrm{i}\}$.

If we consider the attribute implications valid in a given formal context, the following definition of the complete and non-redundant set of attribute implications can be given.

Definition 7. *Let $\langle B, A, R\rangle$ be a formal context and let T_A be a set of attribute implications over A. Then T_A is called complete set of attribute implications in $\langle B, A, R\rangle$ if for an arbitrary $Z \to W$ holds that $Z \to W$ is valid in $\langle B, A, R\rangle$ if and only if $T_A \models Z \to W$. The complete and non-redundant set of attribute implications in $\langle B, A, R\rangle$ is called a basis of $\langle B, A, R\rangle$.*

The main question here is how to test completeness of the set of attribute implications T_A in a formal context and how to find a complete non-redundant set, i. e. basis. The answer to the first question includes testing if the set of all models of T_A equals to the set of all intents of a given formal context. If follows from Definition 4, Lemma 1 and it considers a formal context $\langle \mathrm{Mod}(T_A), A, \ni\rangle$.

	i	ii	iii
a	×		
b	×	×	
c			×

→

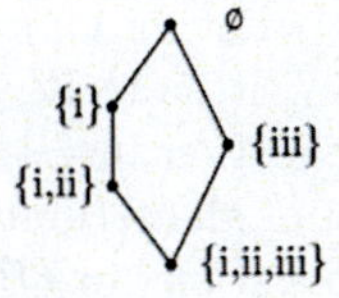

Fig. 3. The set of all intents

The set of all intents of the formal context $\langle B, A, R\rangle$ (Fig. 3) equals to the set $\mathrm{Mod}(\{\mathrm{ii}\} \to \{\mathrm{i}, \mathrm{ii}\}, \{\{\mathrm{i}, \mathrm{iii}\} \to \{\mathrm{i}, \mathrm{ii}, \mathrm{iii}\}) = \{\emptyset, \{\mathrm{i}\}, \{\mathrm{iii}\}, \{\mathrm{i}, \mathrm{ii}\}, \{\mathrm{i}, \mathrm{ii}, \mathrm{iii}\}\}$, hence the set $\{\{\mathrm{ii}\} \to \{\mathrm{i}\}, \{\mathrm{i}, \mathrm{iii}\} \to \{\mathrm{ii}\}\}$ is a complete set of attribute implications in $\langle B, A, R\rangle$.

The second question requires the notion of pseudointent defined recursively as follows.

Definition 8. *Let $\langle B, A, R\rangle$ be a formal context and $Y, Z \subseteq A$. Then Z is called a pseudointent of $\langle B, A, R\rangle$ if $Z \subset Z^{\swarrow\nearrow}$ and $Y^{\swarrow\nearrow} \subseteq Z$ for every pseudointent Y such that $Y \subset Z$.*

The set $T_A = \{Z \to Z^{\swarrow\nearrow} : Z$ is a pseudointent of $\langle B, A, R\rangle\}$ is a basis of $\langle B, A, R\rangle$. For the detailed proofs, see [3,27].

	i	ii	iii
a	×		
b	×	×	
c			×

Z	$Z^{\swarrow\nearrow}$	$Z \subset Z^{\swarrow\nearrow}$	$Y^{\swarrow\nearrow} \subseteq Z$
$\emptyset$	$\emptyset$	no	no
{i}	{i}	no	no
{i,ii}	{i,ii}	no	no
{i,ii,iii}	{i,ii,iii}	no	no
{i,iii}	{i,ii,iii}	yes	yes
{ii}	{i,ii}	yes	yes
{ii,iii}	{i,ii,iii}	yes	no
{iii}	{iii}	no	no

Fig. 4. Testing of being pseudointent in lexicographical order

As we can see in Fig. 4, {ii} and {i, iii} are pseudointents of the given $\langle B, A, R\rangle$, but {ii, iii} is not a pseudointent even though $\{\mathrm{ii}, \mathrm{iii}\} \subseteq \{\mathrm{ii}, \mathrm{iii}\}^{\swarrow\nearrow}$. Remind that the attribute implication $Z \to Z^{\swarrow\nearrow}$ is valid in $Y \subseteq A$ if $Z \not\subseteq Y$ or $Z^{\swarrow\nearrow} \subseteq Y$. But $Z^{\swarrow\nearrow} \subseteq Y$ implies $(Z^{\swarrow\nearrow} \setminus Z) \subseteq Y$, since $(Z^{\swarrow\nearrow} \setminus Z) \subseteq Z^{\swarrow\nearrow}$. On the other hand, let $(Z^{\swarrow\nearrow} \setminus Z) \subseteq Y$. If $Z \subseteq Y$, then $(Z^{\swarrow\nearrow} \setminus Z) \subseteq Y$ implies $Z^{\swarrow\nearrow} \subseteq Y$. The set $T_A = \{Z \to (Z^{\swarrow\nearrow} \setminus Z) : Z$ is a pseudointent of $\langle B, A, R\rangle\}$ is also complete and non-redundant set called Guigues-Duquenne basis of $\langle B, A, R\rangle$ and one can prove that this basis is the smallest one from the all complete set of attribute implications valid in a formal context $\langle B, A, R\rangle$, see for instance [27].

The validity of attribute implication $Z \to W$ in a formal context can be alternatively checked by $Z^{\swarrow} \subseteq W^{\swarrow}$ that comes from the natural requirement that all objects having all attributes from Z have all attributes from W. It can be formally rewritten to $W \subseteq Z^{\swarrow\nearrow}$ (compare with Definition 4). The validity of attribute implication in $\langle B, A, R\rangle$ one can denote by $\langle B, A, R\rangle \models Z \to W$. We denote the set of all attribute implications over A by $\mathrm{Imp}(A)$, the set of all valid attribute implications in $\langle B, A, R\rangle$ by $\mathrm{T}_{\langle B,A,R\rangle}$ and finally, Guigues-Duquenne basis of $\langle B, A, R\rangle$ denote by $\mathrm{Tb}_{\langle B,A,R\rangle}$. We have described the way how to find the minimal set of attribute implications in a formal context which is complete and non-redundant. From the previous reasoning and our running example, we have that $\text{Tb}_{\langle B,A,R\rangle} = \{\{\text{ii}\} \to \{\text{i}\}, \{\text{i}, \text{iii}\} \to \{\text{ii}\}\}$.

In the following, we describe the process of attribute exploration in a formal context, which uses a domain expert.

Definition 9. *Let A be the set of attributes, $Z, Z', W, W' \subseteq A$ and $Y \in \mathcal{P}(A)$. A domain expert on A is a function $e : \text{Imp}(A) \to \mathcal{P}(A) \cup \{\checkmark\}$ such that*

- *either $e(Z \to W) = Y$, whereby $Z \to W$ is not valid in Y*
- *or $e(Z \to W) = \{\checkmark\}$, whereby for all $Z' \to W'$ such that $e(Z' \to W') = Y$ holds $Y \in \text{Mod}(\{Z \to W\})$.*

The set $\{Z \to W : e(Z \to W) = \{\checkmark\}\}$ is denoted by T_e.

In the first case, a user put a counterexample for a given attribute implication. The second case corresponds to the confirmation of the attribute implication by user. Let $A = \{\text{i}, \text{ii}, \text{iii}\}$ and let e_1, e_2 be two domain experts on A (Fig. 5, Fig. 6) such that

a) the first domain expert agrees to all the attribute implication over A, i.e. $e_1(Z \to W) = \{\checkmark\}$ for all $Z \to W \in \text{Imp}(A)$,
b) the second domain expert agrees to all the attribute implication over A except for $\{\text{ii}\} \to \{\text{i}, \text{ii}\}$, more concretely $e_2(\{\text{ii}\} \to \{\text{i}, \text{ii}\}) = \{\text{ii}\}$ and $e_2(Z \to W) = \{\checkmark\}$ for all $Z \to W \in \text{Imp}(A) \setminus \{\{\text{ii}\} \to \{\text{i}, \text{ii}\}\}$.

The algorithm for attribute exploration in a formal context can be described as follows.

Definition 10. *Let $\langle B, A, R\rangle$ be a formal context, let e be a domain expert on A and let $T_A \subseteq T_e \subseteq T_{\langle B,A,R\rangle}$.*

1. *Find the greatest model of T_A (in lexicographical order) and denote as Z.*
2. *If Z is not pseudointent of $\langle B, A, R\rangle$, follow Step 5.*
3. *If $e(Z \to Z^{\swarrow\nearrow}) = \{\checkmark\}$, then $T_A = T_A \cup \{Z \to (Z^{\swarrow\nearrow} \setminus Z)\}$, follow Step 1.*
4. *If $e(Z \to Z^{\swarrow\nearrow}) = Y$, then $B = B \cup \{x\}$ and $\{x\}^{\nearrow} = Y$, follow Step 1.*
5. *Find the greatest model of T_A smaller than Z (in lexicographical order), denote this model as Z and follow Step 2.*

We obviously take $T_A = \emptyset$ at the beginning. With Step 3, we add the new attribute implication into the set T_A and the models of T_A are recalculated. With Step 4, a new object into a formal context is added, T_A does not change, but we switch again to the greatest model of T_A, because $Z^{\swarrow\nearrow}$ has to be recalculated. Note that all attribute implications in T_A remain valid in $\langle B, A, R\rangle$, because of second part of Definition 9. If the greatest model of T_A smaller than Z in lexicographical order does not exist in Step 5, we terminate. One can prove that T_A after termination forms Guigues-Duquenne basis of $\langle B, A, R\rangle$. However, note that set B and relation R can be extended during the calculation. We can omit the computing of models of T_A, if we compute directly the set of all pseudointents

T_A	Z	pseudointent	$Z \to Z^{\swarrow\nearrow}$	e_1
∅	{iii}	no		
	{ii,iii}	no		
	{ii}	yes	{ii} → {i,ii}	confirms
{{ii} → {i}}	{iii}	no		
	{i,iii}	yes	{i,iii} → {i,ii,iii}	confirms
{{ii} → {i}, {i,iii} → {ii}}	{iii}	no		
	{i,ii,iii}	no		
	{i,ii}	no		
	{i}	no		
	∅	no		

Fig. 5. Attribute exploration of expert e_1 in $\langle B, A, R\rangle$

T_A	Z	pseudointent	$Z \to Z^{\swarrow\nearrow}$	e_2
∅	{iii}	no		
	{ii,iii}	no		
	{ii}	yes	{ii} → {i,ii}	counterex. {ii}
∅	{iii}	no		
	{ii,iii}	yes	{ii,iii} → {i,ii,iii}	confirms
{{ii,iii} → {i}}	{iii}	no		
	{ii}	no		
	{i,iii}	yes	{i,iii} → {i,ii,iii}	confirms
	{iii}	no		
{{ii,iii} → {i}, {i,iii} → {ii}}	{ii}	no		
	{i,ii,iii}	no		
	{i,ii}	no		
	{i}	no		
	∅	no		

Fig. 6. Attribute exploration of expert e_2 in $\langle B, A, R\rangle$

of $\langle B, A, R\rangle$ instead of models of T_A. The illustration of attribute exploration process of two domain experts from running example is shown in Fig. 5 and Fig. 6.

Interesting modifications and alternative approaches to the attribute exploration can be found in [28–30] and at the present time, the process of conceptual exploration is investigated by Ganter, Obiedkov, Rudolph and Stumme.

Finally, we present the lemma in order to describe the generation of the attribute implications valid in a formal context from the concept lattice with reduced labeling.

Lemma 2. *Let $\langle B, A, R\rangle$ be a formal context and $Z, W \subseteq A$. Then $Z \to W$ is valid in $\langle B, A, R\rangle$ if and only if* $\inf\{\alpha(z) : z \in Z\} \preceq \inf\{\alpha(w) : w \in W\}$, *whereby* $\alpha(y) = \langle \{y\}^{\swarrow}, \{y\}^{\swarrow\nearrow}\rangle$ *for every* $y \in A$.

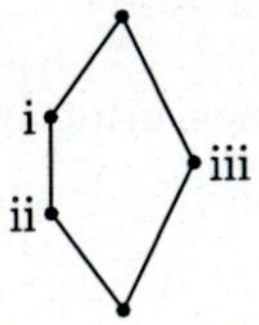

∅	→	∅
{i}	→	∅, {i}
{ii}	→	∅, {i}, {ii}, {i,ii}
{iii}	→	∅, {iii}
{i,ii}	→	∅, {i}, {ii}, {i,ii}
{i,iii}	→	∅, {i}, {ii}, {iii}, {i,ii}, {i,iii}, {ii,iii}, {i,ii,iii}
{ii,iii}	→	∅, {i}, {ii}, {iii}, {i,ii}, {i,iii}, {ii,iii}, {i,ii,iii}
{i,ii,iii}	→	∅, {i}, {ii}, {iii}, {i,ii}, {i,iii}, {ii,iii}, {i,ii,iii}

Fig. 7. The attribute implications valid in $\langle B, A, R\rangle$ from a concept lattice with reduced labeling

Proof. It holds $Z \to W$ is valid in $\langle B, A, R\rangle$ if and only if $Z^{\swarrow} \subseteq W^{\swarrow}$ if and only if $\langle Z^{\swarrow}, Z^{\swarrow\nearrow}\rangle \preceq \langle W^{\swarrow}, W^{\nearrow}\rangle$ if and only if $\inf\{\alpha(z) : z \in Z\} \preceq \inf\{\alpha(w) : w \in W\}$. □

All the attribute implications from Fig. 7 (conclusions of attribute implications are separated by commas, in total 37) follow semantically from $\mathrm{Tb}_{\langle B,A,R\rangle} = \{\{\mathrm{ii}\} \to \{\mathrm{i}\}, \{\mathrm{i}, \mathrm{iii}\} \to \{\mathrm{ii}\}\}$.

The attribute implications in formal concept analysis is connected with association rules proposed by Agrawal et al. [31]. If we will consider association rules in a formal context, then one can define the support as the number of objects which contain attributes from the premise. The confidence one can define as the percentage of objects that contain attributes from the premise of association rule also contain the attributes from the conclusion of the association rule. If confidence equals one, it corresponds to the attribute implication valid in a formal context and we can call it as exact association rule. Otherwise, we will call it partial attribute implication valid in a formal context or approximate association rule [32–34].

Definition 11. *Let $\langle B, A, R\rangle$ be a formal context, $Z, W \subseteq A$ and $Z \cap W = \emptyset$. An association rule $Z \to W$ over A in a formal context $\langle B, A, R\rangle$ is an implication $Z \to W$ over A. The support of an association rule $Z \to W$ over A in a formal context $\langle B, A, R\rangle$ can be defined as*

$$\mathrm{support}(Z \to W) = |Z^{\swarrow}|.$$

The confidence of an association rule $Z \to W$ over A in a formal context $\langle B, A, R\rangle$ can be defined as

$$\mathrm{confidence}(Z \to W) = \frac{|(Z \cup W)^{\swarrow}|}{|B|}.$$

If $\mathrm{confidence}(Z \to W) = 1$, *the association rule $Z \to W$ over A in a formal context $\langle B, A, R\rangle$ is also called the exact association rule (and it corresponds to the attribute implication in a formal context).*

If confidence$(Z \to W) < 1$, *the association rule* $Z \to W$ *over* A *in a formal context* $\langle B, A, R\rangle$ *is also called the approximate association rule in a formal context (or it is called the partial attribute implication in a formal context).*

We will apply the notion of partial attribute implication for measuring of pupils' computational thinking in Sect. 4.

3 Example of Attribute Exploration

In formal concept analysis, the mappings act on different types of structures, e.g. chains, partially ordered sets, complete lattices, Boolean lattices, distance lattices and other structures. We remind the definitions of the several structures used in this section:

Definition 12. *A binary relation* R *that is reflexive, antisymmetric, transitive is called a partial order. A partial order will be denoted as* $\preceq$. *An example for a partial order* $\preceq$ *is a relation* $\leq$. *A set* P *equipped with a partial order* $\preceq$ *is called a partially ordered set (or shortly poset).*

Definition 13. *A partially ordered set* $(P, \preceq)$ *is called a complete lattice (shortly denoted as* L*) if every subset* S *of* P *has the greatest lower bound (called the infimum or join) and the least upper bound (called the supremum or join). The elements* $\inf S$ *and* $\sup S$ *can be alternatively characterized by*

$$(\forall p \in P)\big((\forall s \in S)s \preceq p \leftrightarrow \inf S \preceq p\big)$$

and

$$(\forall p \in P)\big((\forall s \in S)p \preceq s \leftrightarrow p \preceq \sup S\big).$$

For dealing with indexed elements one can also use notation $\bigwedge_{i\in I}$ *or* $\bigvee_{i\in I}$ *for infimum and supremum, respectively.*

Partially ordered sets and complete lattices are the well known generalizations of linearly ordered sets (also called totally ordered sets or chains, for instance the set of natural numbers equipped with $\leq$ ordering). Posets and complete lattices can concern with incomparable elements and may be illustrated by Hasse diagram. We enclose Fig. 8 that intercepts the differences between these structures.

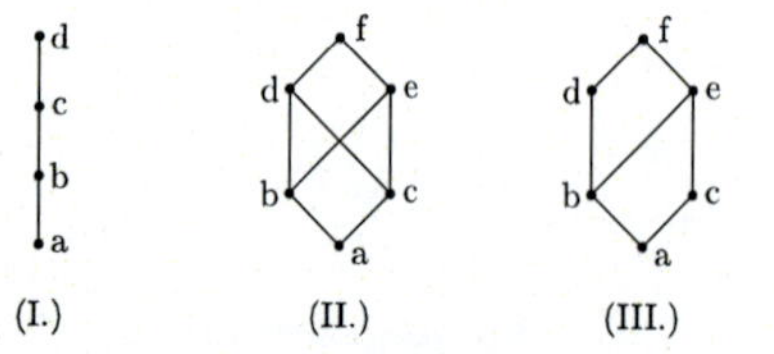

Fig. 8. Finite chain, poset and complete lattice

Every chain is a poset and every finite chain is a complete lattice. Every complete lattice is a poset. In contrary, it does not hold that every poset is a complete lattice, since the supremum of subset $\{b, c\}$ of poset (II.) does not exist. Taking into account the infinite chains, the infinite set of all integers has not a supremum and the infinite subset of rational numbers q such that $q^2 < 3$ has neither supremum nor infimum.

From a theoretical point of view, the distributive lattices are the most thoroughly studied in the literature.

Definition 14. *A complete lattice L is called distributive if for every $x, y, z \in L$*

$$\inf\{x, \sup\{y, z\}\} = \sup\{\inf\{x, y\}, \inf\{x, z\}\}$$

or equivalently

$$\sup\{x, \inf\{y, z\}\} = \inf\{\sup\{x, y\}, \sup\{x, z\}\}.$$

The complete lattice (III.) from Fig. 8 is distributive, the complete lattices (I.) and (II.) are not distributive – take their elements c, d, e. The powerset of an arbitrary set equipped with a partial order of subset relation is an important and special case of a complete lattice.

Remark 1. Let X be a set and $\mathcal{P}(X)$ be the powerset of X. Then $(\mathcal{P}(X), \subseteq)$ is a complete lattice with $\bigwedge_{i\in I} A_i = \bigcap_{i\in I} A_i$ and $\bigvee_{i\in I} A_i = \bigcup_{i\in I} A_i$ such that $A_i \in \mathcal{P}(X)$ for $i \in I$. (The existence of all such unions and intersections in $\mathcal{P}(X)$ guarantees the existence of supremum and infimum.)

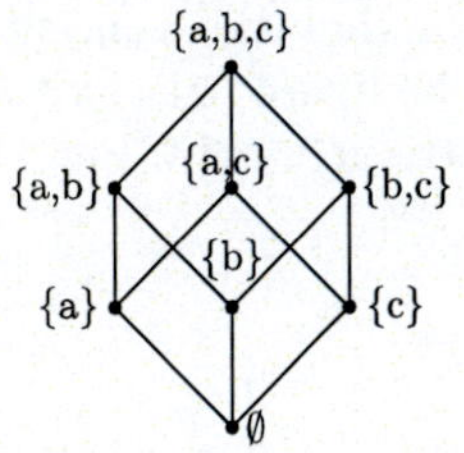

Fig. 9. Complete lattice of a powerset of $X = \{a, b, c\}$

In an arbitrary complete lattice L, the existence of $\inf \emptyset$ guarantees the greatest element 1_L in L and $\sup \emptyset$ guarantees the least element 0_L in L. The well known complete lattices are complete Boolean lattices which can be also viewed as a Boolean algebra with the operations of supremum, infimum and complement.

Definition 15. *A complete lattice L is called complemented if each element x has a complement x' such that $\inf\{x, x'\} = 0_L$ and $\sup\{x, x'\} = 1_L$. A complete Boolean lattice is a complete lattice that is complemented and distributive.*

	chain	poset	comp_lat	dist_lat	Bool_lat
	×	×	×	×	
		×			
		×	×	×	
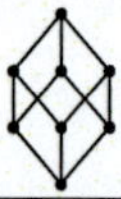		×	×	×	×

Fig. 10. A formal context before the attribute exploration

Note that the complete lattice of powerset X from the Fig. 9 is a complete Boolean lattice.

Since the process of attribute exploration provides a possibility to ask about the validity of attribute implications, we illustrate the example of attribute exploration on the special object-attribute table. As objects consider the specific examples of partially ordered sets and for the attributes take their membership to the general types of partially ordered sets. At the beginning, we take four specific examples and their properties as shown in Fig. 10.

In Sect. 3, we have shown that the concept lattice corresponding to the explored formal context can be figured also with a reduced labeling, especially with a reduced labeling of attributes only (Fig. 11).

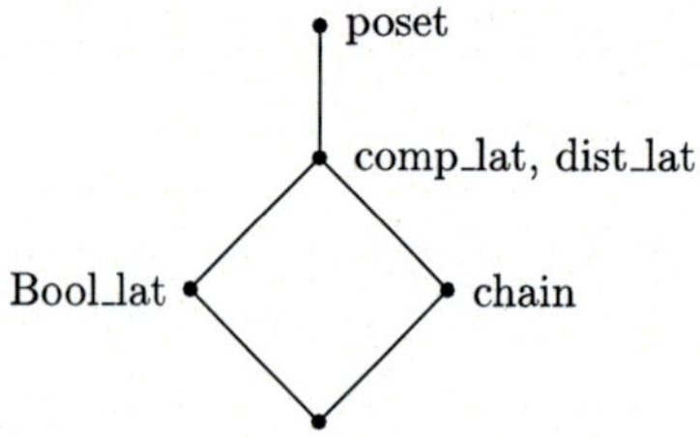

Fig. 11. Concept lattice with reduced labeling of attributes

We can read the attribute implications from this concept lattice. For instance, we can see that the type of Boolean lattices implies the type of complete lattices. Another attribute implication expresses that the type of chains implies the type of complete lattices. However, some counterexample to this attribute implication

No.	question	answer	given counterexample
1.	{poset,Bool_lat} → {comp_lat,dist_lat}?	yes	
2.	{poset,dist_lat} → {comp_lat}?	yes	
3.	{poset,comp_lat} → {dist_lat}?	no	{poset,comp_lat}
4.	{chain,poset} → {comp_lat,dist_lat}?	no	{chain, poset}
5.	{chain,poset,comp_lat} → {dist_lat}?	yes	
6.	∅ → {poset}?	yes	

Fig. 12. A table of attribute exploration

	chain	poset	comp_lat	dist_lat	Bool_lat
	×	×	×	×	
		×			
		×	×	×	
		×	×	×	×
		×	×		
	×	×			

Fig. 13. A formal context after the attribute exploration

can be found, it means that Fig. 10 does not contain all possible cases. To correct this, we can control the validity of attribute implications from Fig. 10 by the process of attribute exploration presented in Sect. 3.

If we take the formal context from Fig. 10, the Guigues-Duquenne basis includes six attribute implications. Therefore, the six questions and answers in Fig. 12 are possible.

By the process of attribute exploration (Fig. 12), we have found two counterexamples which allow us to classify the various types of structures even more precise. If we take the counterexamples of attribute implications answered by no (this means No. 3, 4), we can append the new rows in our original formal context as shown in Fig. 13.

And finally, the concept lattice can be recalculated after an addition of the first counterexample (left part of Fig. 14) and after an appending of the second

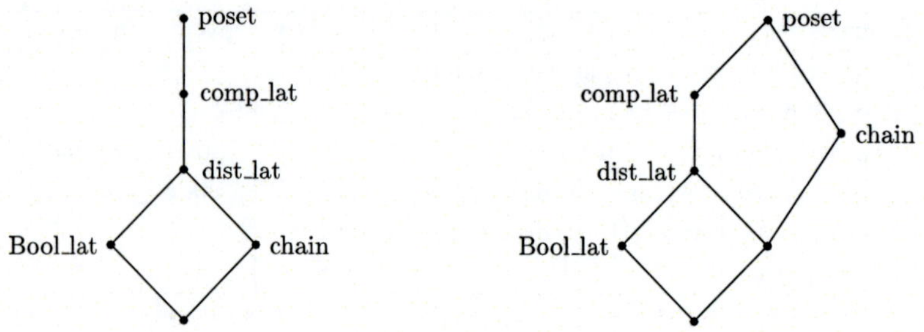

Fig. 14. Concept lattices after the first and the second counterexample

counterexample (right part of Fig. 14). The second concept lattice shows the natural expectation that not every chain is a complete lattice and also that not every complete lattice is a distributive lattice (which was not included in the input formal context from Fig. 10).

4 Measuring of Pupils' Computational Thinking

In this section, we focus on measuring the level of pupils' computational thinking, which according to [35] can be described as all about solving problems effectively – with or without using of the computer. Computational thinking is a complex ability made up of several concepts. Several major educational companies (e.g. Computer Science Teachers Association, International Society for Technology in Education, Computing at School) have designed their own frameworks for computational thinking concepts. For our research purposes, we have adopted six concepts proposed by Computing at School [35], namely:

1. Logic – predict, analyze (logically draw conclusions from observations and experiments, analyze a system/program and predict its behavior).
2. Algorithms – create steps and rules (compile activity procedures for an executor (algorithms, programs, scenarios, storyboards) and also use, modify, improve the created procedures (algorithms, programs, scenarios, storyboards).
3. Decomposition – division into parts (divide the problem into easier subproblems, the solution of which will be usable in solving the original problem).
4. Pattern search – recognize and use similarities (recognize and determine the same/similar parts/properties/rules in the system and use the found patterns in different activities/problem solving).
5. Abstraction – choose the essential, disregard the less substantial (determine which details/elements/properties/relationships of the system are essential in a given situation and which we can neglect).
6. Evaluation – making decisions (determine relevant evaluation criteria and evaluate project/program/algorithm according to them).

For measuring of computational thinking we have developed own test. At the beginning of the development of the computational thinking test, we set several requirements, which include mainly: independence from specific subject matter, coverage for all age groups of pupils (ages 11–19), administrability of the test within one lesson and the possibility of machine evaluation of the test solution. The computational thinking test is a test of a pupil's relative performance (norm-referenced test). We evaluate the pupil's success within his category (age, type of school) and express it with a percentile.

We developed the own test for measuring of computational thinking in several consecutive stages. From a bank of 30 candidate assignments, we compiled a printed beta version of the test with 2 closed and 10 open assignments, which was administered in November 2017 with a group of 26 pupils of the extended study of computer science. Subsequently, we created a pilot version of the test with 8 closed and 3 open items, which we administered to 702 participants of the target group in December 2017 to January 2018. Finally, we created the final version of the test with 10 closed and 2 open items and 4 attitude questions. We administered the final test from March 2018 to June 2020.

We present the characteristics of objects in our data set in Table 1.

Table 1. Characteristics of objects in data set

Characteristic	Count
Pupils with pre-test	29,461
Pupils with pre-test only	20,343
Pupils with post-test	9,118
Pupils with pre- and post-test	9,118
Pupils with pre- and post-test studying at least one special study material	1,357
Pupils with pre- and post-test passing at least one Python study material	142

Within the data set, we explored 56 attributes regarding the pupils with pre-test and post-test. In total, 39 attributes were analyzed for pupils with pre-test only. Nominal attributes include gender, organization subjective assessment of programming interest by pupils, subjective assessment of programming importance by pupils, subjective assessment of programming difficulty by pupils, gender, organization. For each of 17 types of special study materials, we have analyzed the number of studied materials between the pre-test and post-test. Moreover, the attribute of the number of rounds in the PALMA junior programming competition (https://di.ics.upjs.sk/palmaj/) is available. The summary results of pre-test and post-tests are given in percentages. In addition, the partial results in 6 concepts proposed by Computing at School [35] for pre-test and post-test are given in percentages. Other attributes include e.g. the number of valid answers, the number of answers "I do not know", the number of missing answers, or the total duration of the test for both pre-test and post-tests.

We applied our own implementations of algorithms in formal concept analysis and Concept Explorer software tool. Moreover, we have used also the implementations in software tools FCART, Lattice Miner or FcaStone in order to explore the advantages and disadvantages of various sources.

We observed the increase in the percentage values in all components of the post-test compared to the pretest, also in the number of valid answers. On the contrary, the number of answers "I don't know", the missing answers and the total filling time in the post-test was shortened compared to the pretest. A total of 5586 out of 9118 pupils improved their result in the final test (61.3%). Out of the total number of 9118 pupils who passed the final test, 1209 pupils studied at least one of the special study materials between pre-test and post-test. From the pupils who studied at least one study material, 793 pupils (65.6%) improved their post-test and 416 worsened (34.4%).

Table 2. Partial attribute implications of 29,461 pupils with pre-test

Partial attribute implications	Support	Confidence
{interesting, difficult} → {important}	9,447	0.77
{interesting} → {important}	13,633	0.75
{important} → {difficult}	14,024	0.73
⋮	⋮	⋮

Regarding all pupils with pre-test, we have generated the partial attribute implications on a set of three attributes: programming is interesting (1 – yes, 0 – no or do not know), programming is important (1 – yes, 0 – no or do not know), programming is difficult (1 – yes, 0 – no or do not know). The results are shown in Table 2.

For the pupils without post-test, we have generated the partial attribute implications on a set of four attributes: programming is not interesting, programming is not important, programming is not difficult, and the total score less than 50%. The results are shown in Table 3.

Table 3. Partial attribute implications of 9,118 pupils without post-test

Partial attribute implications	Support	Confidence
{not inter., not import., not diffic.} → {score < 50%}	1,669	0.83
{not important} → {score < 50%}	6,976	0.75
{not difficult} → {score < 50%}	6,370	0.72
⋮	⋮	⋮

Regarding the pupils who passed the pre-test and post-test, we analyzed the gender, summary improvement, improvement in 6 concepts (algor, eval, abstract, decomp, pattern, logic) and studying some of the study materials between pre-test and post-test. We found the partial attribute implications which are shown in Table 4.

Table 4. Partial attribute implications of 9,118 pupils with pre-test and post-test

Partial attribute implications	Support	Confidence
{algor, eval, abstract, decomp} → {improvement}	1,156	1
{boy, algor, eval, decomp} → {improvement}	1,029	0.99
{algor, eval} → {improvement}	2,893	0.96
{eval} → {improvement}	5,155	0.85
{algor, eval, materials} → {improvement}	459	0.96
{boy, algor, abstract, materials} → {improvement}	152	0.96
⋮	⋮	⋮

For the pupils who passed the pre-test and post-test and have studied at least one special study material between pre-test and post-test, we analyzed the gender, summary improvement, and 16 types of study materials (Scratch, BBC micro:bit, 3D print, Video, Lego, Coding, Geolocation, Ethical aspects, E-mail, Risks of ICT, Python, Information safety, Artificial intelligence, Raspberry, AppInventor, Presentation) and participation at PALMA junior programming competition. In Table 5, we present several partial attribute implications for this group of pupils.

Table 5. Partial attribute implications of 1,357 pupils with pre-test, post-test and studying special materials between pre-test and post-test

Partial attribute implications	Support	Confidence
{Scratch, Geolocation} → {boy}	22	0.91
{Lego} → {improvement}	55	0.8
{BBC micro:bit} → {improvement}	160	0.77
{PALMA junior} → {improvement}	43	0.74
{Scratch} → {improvement}	218	0.74
{Python} → {improvement}	142	0.62
⋮	⋮	⋮

Finally, we have explored 142 pupils with pre-test, post-test and studying special Python study materials between pre-test and post-test. We analyzed gender, age (1 – more than 17 years), the improvement, the time of post-test (1

– more than 25 min), the number of study materials (1 – more than 3 materials), programming is interesting (1 – yes), programming is important (1 – yes), programming is difficult (1 – yes), norm summary improvement more than 20% $\left(1 \text{ if } \frac{\text{post-test}-\text{pre-test}}{\text{max}-\text{pre-test}} > 0.2\right)$. In Table 6, we present several partial attribute implications for this group of pupils.

Table 6. Partial attribute implications of 142 pupils with pre-test, post-test and studying special Python materials between pre-test and post-test

Partial attribute implications	Support	Confidence
{age, materials, interest, improvement} → {norm0.2}	35	0.91
{age} → {importance}	111	0.82
{age, time, improvement} → {norm0.2}	40	0.78
{boy, interest} → {improvement}	41	0.73
{time, interest} → {improvement}	45	0.71
⋮	⋮	⋮

5 Conclusion

We presented the computational methods for generation of attribute implications in the field of formal concept analysis and illustrated the process of attribute exploration for special types of object-attributes tables. We proposed the own test for measuring of pupils' computational thinking in several consecutive stages and applied the partial attribute implications in formal concept analysis on the results of national testing of pupils in the field of computer thinking.

We have found that 83% of pupils who considered programming as not interesting, not important and not difficult, obtained the total scores in pre-test less than 50%. In contrary, improvements in algorithmic, evaluation, abstract and decomposition concepts imply the total improvement. We explored the types of study materials between pre-test and post-test, which imply the total improvement. Regarding the special Python materials between pre-test and post-test, the attributes including age more than 17 years, studying more than three Python study materials, interest in programming and total improvement implied the norm total improvement more than 20%.

We are convinced that the area of attribute implications plays an important role in the research of both theoreticians and practitioners and thus, the research from a broader perspective is still beneficial. In our research, we have prepared the first prospects in some other areas, which we do not introduce in this paper. However, they serve the promising fields of our future examination, hereby we remind for example the theory of rough sets and the approximation operators, the factorization of a formal context and ternary relations, aggregations on the bounded lattices or multilattices, interval and reciprocal relations in a formal context, the cooperative game theory with players and their alternatives, knowledge space theory, or category theory.

Acknowledgement. This work was partially supported by the Slovak Research and Development Agency under the contract No. APVV-15-0091 Effective algorithms, automata and data structures.

We would like to thank the two anonymous reviewers for their suggestions and comments. Following the suggestions, we included several improvements in the chapter. We can provide anonymized dataset up to private request of person interested.

References

1. Wille, R.: Restructuring lattice theory: an approach based on hierarchies of concept. In: Ordered Sets, pp. 445–447. Reidel, Dodrecht-Boston (1982)
2. Kuznetsov, S.O.: Machine learning on the basis of formal concept analysis. Autom. Remote Control **62**(10), 1543–1564 (2001)
3. Ganter, B., Wille, R.: Formal Concept Analysis, Mathematical Foundation. Springer, Heidelberg (1999)
4. Ganter, B., Rudolph, S., Stumme, G.: Explaining data with formal concept analysis. In: Reasoning Web. Explainable Artificial Intelligence, pp. 153–195. Springer, Cham (2019)
5. Ferré, S., Huchard, M., Kaytoue, M., Kuznetsov, S.O., Napoli, A.: Formal concept analysis: from knowledge discovery to knowledge processing. In: A Guided Tour of Artificial Intelligence Research, pp. 411–445. Springer, Cham (2020)
6. Bělohlávek, R., Outrata, J., Trnečka, M.: Toward quality assessment of Boolean matrix factorizations. Inf. Sci. **459**, 71–85 (2018)
7. Ganter, B., Obiedkov, S.: Concept lattices. In: Conceptual Exploration, pp. 1–38. Springer, Heidelberg (2016)
8. Antoni, Ľ., Krajči, S., Krídlo, O., Macek, B., Pisková, L.: On heterogeneous formal contexts. Fuzzy Set. Syst. **234**, 22–33 (2014)
9. Krídlo, O., Krajči, S., Antoni, Ľ.: Formal concept analysis of higher order. In: Ojeda-Aciego, M., Outrata, J. (eds.) International Conference on Concept Lattices and Applications, CLA 2013, pp. 117–128. Laboratory L3i, University of La Rochelle (2013)
10. Krídlo, O., Krajči, S., Antoni, Ľ.: Formal Concept Analysis of higher order. Int. J. Gen. Syst. **45**(2), 116–134 (2016)
11. Antoni, Ľ., Guniš, J., Krajči, S., Krídlo, O., Šnajder, Ľ.: The educational tasks and objectives system within a formal context. In: CLA 2014, pp. 35–46. UPJŠ in Košice, Košice (2014)
12. Krídlo, O., Ojeda-Aciego, M.: Revising the link between L-Chu correspondences and completely lattice L-ordered sets. Ann. Math. Artif. Intell. **72**(1–2), 91–113 (2014)
13. Krídlo, O., Konečný, J.: On biconcepts in formal fuzzy concept analysis. Inf. Sci. **375**, 16–29 (2017)
14. Antoni, Ľ., Cabrera, I.L., Krajči, S., Krídlo, O., Ojeda-Aciego, M.: The Chu construction and generalized formal concept analysis. Int. J. Gen. Syst. **46**(5), 458–474 (2017)
15. Heunen, C.: Categorical Quantum Models and Logics. Amsterdam University Press, Amsterdam (2009)

16. Carpineto, C., Romano, G.: Concept Data Analysis. Theory and Applications. Wiley, Hoboken (2004)
17. Ganter, B., Stumme, G., Wille, R. (eds.): Formal Concept Analysis: Foundations and Applications, vol. 3626. Springer, Heidelberg (2005)
18. Poelmans, J., Ignatov, D.I., Kuznetsov, S.O., Dedene, G.: Formal concept analysis in knowledge processing: a survey on applications. Expert Syst. Appl. **40**(16), 6538–6560 (2013)
19. Priss, U., Riegler, P., Jensen, N.: Using FCA for modelling conceptual difficulties in learning processes. In: Contributions to the 10th International Conference ICFCA 2012, Leuven, Belgium, pp. 161–173 (2012)
20. Poelmans, J., Dedene, G., Verheyden, G., Van Der Mussele, H., Viaene, S., Peters, E.: Combining business process and data discovery techniques for analyzing and improving integrated care pathways. In: The 10th IEEE International Conference on Data Mining 2010, pp. 505–517. Springer, Heidelberg (2010)
21. Kiseliová, T., Krajči, S.: Generation of representative symptoms based on fuzzy concept lattices. Adv. Soft. Comput. **2**, 349–354 (2005)
22. Bělohlávek, R., Sklenář, V., Zacpal, J., Sigmund, E.: Evaluation of questionnaires by means of formal concept analysis. In: Proceedings of CLA 2007, Montpellier, France, pp. 100–111 (2007)
23. Bělohlávek, R., Sigmund, E., Zacpal, J.: Evaluation of IPAQ questionnaires supported by formal concept analysis. Inf. Sci. **181**(10), 1774–1786 (2011)
24. Horák, Z., Kudelka, M., Snášel, V.: FCA as a tool for inaccuracy detection in content-based image analysis. In: Proceedings of the 2010 IEEE International Conference on Granular Computing, pp. 223–228 (2010)
25. Alcalde, C., Burusco, A., Fuentes-González, R.: Application of the L-fuzzy concept analysis in the morphological image and signal processing. Ann. Math. Artif. Intell. **72**, 115–128 (2014)
26. Bělohlávek, R., Vychodil, V.: Attribute implications in a fuzzy setting. In: Formal Concept Analysis, pp. 45–60. Springer, Heidelberg (2006)
27. Bělohlávek, R.: Introduction to Formal Concept Analysis. Palacký University, Olomouc (2008)
28. Ganter, B.: Attribute exploration with background knowledge. Theor. Comput. Sci. **217**(2), 215–233 (1999)
29. Stumme, G.: Attribute exploration with background implications and exceptions. In: Data Analysis and Information Systems. Studies in Classification, Data Analysis, and Knowledge Organization, pp. 457–469. Springer, Heidelberg (1996)
30. Borchmann, D.: A general form of attribute exploration. In: LTCS-Report 13-02, TU Dresden, 17 p. (2013)
31. Agrawal, R., Imielinski, T., Swami, A.: Mining association rules between sets of items in large databases. In Proceedings of the ACM SIGMOD Conference on Management of Data, pp. 207–216 (1993)
32. Luxenburger, M.: Implications partielles dans un contexte. Mathématiques, Informatique et Sciences Humaines **29**(113), 35–55 (1991)
33. Stumme, G., Taouil, R., Bastide, Y., Pasquier, N., Lakhal, L.: Intelligent structuring and reducing of association rules with formal concept analysis. In: Annual Conference on Artificial Intelligence, pp. 335–350 (2001)
34. Lakhal, L., Stumme, G.: Efficient mining of association rules based on formal concept analysis. In: Formal Concept Analysis, pp. 180–195 (2005)
35. BCS Barefoot: Computational Thinking Concepts and Approaches. https://www.barefootcomputing.org/concept-approaches/computational-thinking-concepts-and-approaches

Flower Master Index for Relational Database Selection and Joining

Michal Kvet(✉) and Karol Matiaško

Žilinská univerzita v Žiline Fakulta riadenia a informatiky, 01026 Žilina, Slovakia
Michal.Kvet@fri.uniza.sk

Abstract. Current relational database systems are forming the main part of the intelligent information and support system. Data inside are secured, protected by the relational integrity covered by the transactions. Access and management are done by the relational algebra. Significant aspects expressing the quality and performance are the data access techniques themselves. The index is one of the key features ensuring data retrieval performance, however, from the other operations point of view, it brings additional costs. If the suitable index for the query is not present, the whole table must be scanned sequentially resulting in poor performance. If the data fragmentation is present, the problem is even deeper. This paper aims to propose its own techniques based on relevant block identification objects stored in private or shared memory areas. It uses rebalancing methods on the extent granularity. If the tables are to be joined and both indexes are not present, usually, the Hash join method is used. The technique of two-dimensional index mapper is discussed in the paper, as well.

Keywords: Database performance · Data access · Indexing · Fragmentation

1 Introduction

Data management forms a significant part of the information processing. Current information systems need reliable, secure, and performance effective solutions for dealing with complex data. The main advantage of the relational paradigm in comparison with other architectures, currently mostly based on non-relational form, is its *structure* and *integrity* covered by the *transactions*. The relational database paradigm is based on the *relations* consisting of the attributes covered by the integrity rules. Between individual relations (*entities*), *relationships* can be identified. Each relation is stored in one *tablespace*. Data themselves are formed in the shape of individual blocks associated with just one tablespace. They can be, however, spread across multiple files on the disc storage [1, 2].

The database system consists of the *instance* on the one side and *database* on the second. The instance is delimited either by the *memory structures*, but by the *background processes* supervising the memory and transferring data, as well. In the *nomount* startup step, the *instance* is defined. Its parameter characteristics can be found in the parameter

J. Paralič et al. (Eds.): DISA 2020, AISC 1281, pp. 181–202, 2021.
https://doi.org/10.1007/978-3-030-63872-6_9

file of the system (*pfile*), however, most of them have a default value, except for the *database name*. When the user executes and request, it is processed by the *background processes* and data are manipulated in the *instance memory*. Thus, data must be loaded from the physical storage for the evaluation and consecutive processing, forming the result set. By *mounting* the database during the startup, *controlfile* describing database structure is loaded and data headers are interconnected with the instance. Finally, after *opening* the database, all data from the database are accessible. For the user process on the client-side, the server process on the server is created. If dedicated server mode is used, 1:1 mapping is defined. For the manipulation operations (DML – *Insert*, *Update*, *Delete* and *Select*), it is inevitable to ensure a transparent, reliable, and performance effective process of loading data supervised by background processes, mostly based on the defined server process for the client. Between the database and instance, data flow can be identified. Persistently, data are stored in the physical storage (database), for the processing, data tuples are loaded. Its performance is a crucial part of the whole database processing.

This paper aims to provide a robust solution for dealing and accessing data. Data rows (tuples) can be obtained by scanning the whole table ecosystem, by scanning all the blocks associated with the table (*Table Access Full Method – TAF*). On the second side of the corridor, there are indexes for direct data location. As a consequence, if the suitable index is present, it will be used by the assumption, that the time consumption will be significantly lowered in comparison with the whole table scanning. The final decision is up to the *database optimizer* evaluating the *statistics* and required *data conditions*. However, what if there is no suitable index present in the system? Currently, the whole table is scanned sequentially (*TAF*), block by block, which is not so perfect promptly. Namely, each block must be transferred from the disc storage into the memory, where it is *parsed*, *evaluated* by *extracting* relevant rows forming result set. Such activity is, however, very expensive. In the ideal (but mostly theoretical) environment, each block is full, thus the number of blocks for the table is optimal. In that case, loading is straightforward and without index, there is no room for improvement. Vice versa, data are dynamic, new tuples are loaded, historical data with no actual information are removed, existing rows are updated consequencing in defining data fragmentations over the blocks [2, 6, 13]. Although there are techniques for shrinking the data blocks [7], vacuuming empty blocks, and attempts for removing block fragments, in practice, there are not used very often [7]. The reason is mostly based on two factors. First of all, there is an assumption, that the data blocks will be necessary for the near future as the result of dynamic data changes [8]. The second aspect is based on the resource consumption [10].

As introduced in this section, data fragmentation and tuple distribution across the data blocks can be the huge performance problem and the bottleneck of the whole system. This paper aims to deal with the performance from the data access point of view. Effective data locating process is a key factor influencing performance [9]. We summarize existing approaches and highlight our proposed techniques to improve performance and limit the fragmentation in the block and extent granularity [10].

This structure is as follows: Sect. 2 deals with the database system architecture and instance management and physical storage. Index access techniques are described in Sect. 3 consequencing in proposing own Master index architecture and access models defined in Sect. 4. In those sections, we deal with the relational algebra operations selection and projection. In the next part, we are treating the table joining methods. Section 5 defines the state of the art in this field, whereas in Sect. 6, our proposed solution for effective table joining is proposed. It is based on the two factors – performance effectivity of the data access followed by the joining operation itself. In Sect. 7, we summarize the reached results in the evaluation environment and executed experiments.

2 Database System Architecture, Instance

The database system is a complex environment consisting of the database *instance* and physical storage operated by the *database*. In this section, we will describe the core principles and structure of the database instance. We will mostly highlight structures playing a role in our proposed solution improving data access operation performance.

Database instance can be characterized by background processes managing database and individual operations required either from the user side, but as the autonomous operations of the database system management, as well. Figure 1 shows the core model of the DBS Oracle architecture. The *user process* on the client-side is the connection initiator. It contacts the database *listener* on the defined *port*, which serves the client connection to the *server process* on the server-side provided by the *Process Monitor* background process. If the interconnection is done, communication between client and server can be straightforward. As evident from Fig. 1, each server process has its own allocated memory region. It is created automatically with the creation of the server process itself in the non-sharable memory. It is used to process *SQL statements* and to cover the logon and other *session information.* SQL statement management region is mostly used for cursor management, local variables storage, sorts, etc. Unlike *PGA* (*Private Global Area*), there is *SGA* (*Shared Global Area),* which is allocated during the database system no-mounting process and it is shared across the whole instance and all connections. It is defined by the following memory structures, the first three (A, B, C) are mandatory, the rest are optional.

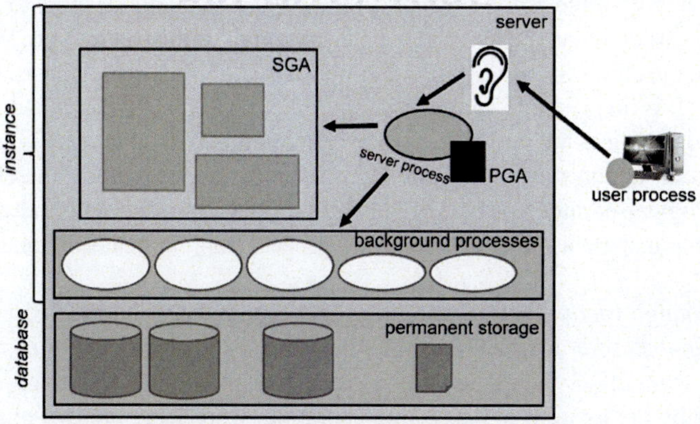

Fig. 1. Database system architecture

A – Buffer cache

The *buffer ca*che is a working area for the SQL statements. Users cannot operate the physical database, it is done by the background processes and individual data operations are always done in the memory and consecutively copied to the physical storage. Thus, if any data portion is to be changed (updated, deleted, inserted), it is done in the memory blocks secured by the transaction logs, therefore no change can be lost, at all. Similarly, if there is a request to obtain data from the database, in the primary phase, particular data must be copied to the memory for the evaluation. Data stored in the database are defined in the block structure, its size is mostly 8 KB (default), but can differ based on the definition during the database system installation. Afterward, it cannot be changed at all.

Individual blocks in the memory *Buffer cache* can be either *clean* (consisting of the data, however, the block content is the same as the image in the physical database), *dirty* (data in the memory do not conform the image in the database, there is at least one bit changed) or *empty* (never used). *Empty* and *clean* buffer block can be used for new data anytime, unlike *dirty* buffer block, which must be copied and applied in the database before rewritten. The loading process is a process of transferring data using the I/O operations from the database to the instance memory and vice versa. It is the most demanding process and it mostly influences the performance of the query processing and also reflects system-wide performance and throughput. In our proposed solution, we deal with the whole loading and locating process to ensure a reliable solution with minimal resource and time consumption necessity. The aim is always to limit the amount of data transferred.

B – Log Buffer

The crucial part ensuring *consistency* and *transaction* management is covered by the *Log buffer*. It is a small, short term staging area used for storing *transaction change vectors* before they are written to the *Redo log* disc storage files. It guarantees no data can be lost. It is associated with the *Undo segment* in the physical storage. It aims to allow object image reconstruction, as it existed at the starting point of the transaction. When queries are defined, the aim is to locate particular relevant data from the database as effectively, as possible. It is, in principle correct assumption, however, such block and tuples inside do not need to reflect the current version based on the *SCN* (*System Change Number)* delimiting transaction start time point and need to be reconstructed. Therefore, in [19], there is a logical pointer layer to the historical images in the temporal environment. Transaction transparency and local recovery process improvements are managed in [10, 22, 25].

Each *change vector* (before and after data image) must be created and stored in the *Log buffer* before any change to the data can occur. It is mostly ensured by two protocols – *direct* [8, 14] and *two-phase* [13, 17]. In direct protocol mode, each change is preceded by the logging as the atomic operation, whereas in the two-phase protocol, all data manipulation is logged in the first phase, then, all changes are applied to the data blocks in the database. One way or another, data must be always logged before the change itself. The log buffer is therefore rightly considered as the performance bottleneck of the system [9, 13]. Storing log data to the database storage is just one side of the problem, which can be partially improved by the parallelism among transactions [28]. The second aspect is just the image composition covering the database and the particular change vector image [4].

C – Shared pool

The shared pool is the most complex structure consisting of many sub-elements. From the evaluation and query processing point of view, the most important structure is the *Library cache.* It is the memory area for the recently executed code in the parsed form – by converting user requests to the executable form by describing individual data access techniques, joining, etc. If the query is already parsed, the database optimizer instructs the execution processes to use such form. If not, the optimizer must choose the best execution plan based on the statistics and heuristical approaches. Other structures are *Data Dictionary Cache* storing recently used object definitions, *PL/SQL area* for storing PL/SQL code. Result cache memory structure can store data produced by the query. If executed multiple times, it is not necessary to obtain data from the database, but if available, they can be located in the memory directly. The database system ensures, that the *Result cache* memory structure is reliable – any change on the data invalidates the stored results, thus it is ensured, that correct data are always produced.

Optional structures include *Large pool*, *Java pool,* and *Streams pool* relevant for the used architecture [8].

2.1 Background Processes

Background processes are part of the system resources managing instance and the database. They play an important role in the process of creating and opening an instance, as well as in supervising the existing system. *System Monitor* main task is associated with the *mount* and *opening* process but has several housekeeping tasks of the *open* instance. *Process Monitor* launches, manages, and deallocates server processes and detects any problems with sessions. When dealing with the performance, the *Database Writer* process is inevitable. *The user process*, nor the *server process*, can manipulate the data directly. *Database Writer* is the process, which is responsible for the writing dirty blocks of the instance memory into the database. Whereas the transfer consists of the expensive I/O operation, the system prefers to minimize the amount of block writing. We reflect such requirements and propose our techniques to minimize data about to be transferred. Note, that data can be fragmented among the database storage, thus the requirement is relevant and significant. In principle, data are written to the disc only if there are no empty blocks in the memory *Buffer cache* to serve new data, although some small data portion is written each three seconds by default [8]. Thus, in this manner, if the whole database and support structures can be placed in the memory, an almost optimal solution can be reached. Unfortunately, current databases are mostly too large to fit the memory size [8]. Moreover, if no block is written to the disc, restore and recovering processes after instance failure would last too much significantly overlapping the *service level agreement (SLA).*

Log Writer operates *Log buffer* and physical log files in the database. It must always precede transaction approval (*commit*) to ensure durability.

In addition to these processes, there is a whole spectrum of background processes covering the *Checkpoint process, Manageability Monitor, Memory Manager, Archiver*, etc. [8, 26].

2.2 Physical Structures

Three file structure types are making up the Oracle database (*controlfile, data files,* and *log files*). *Controlfile* is a root element consisting of the pointers to the database structures – individual data and log files, as well as many parameters forming a database. It stores critical information declaring integrity, if the instance fails, like *SCN*, timestamps, etc. [31]. Change vectors are stored in the *Online redo log* file category. The data repository is created by the data files encapsulated in the tablespaces [27]. Each tablespace consists of the parameter specifying data block size, most often 8 KB block is used. *DB_BLOCK* initialization parameter specifies the structure among the tablespace, specification of the Buffer cache structure is influenced by such parameter, as well. Evaluation of the data block size to the I/O operation performance can be found in [24]. The tablespace can be formed by multiple files covering multiple data objects. Vice versa, each data file can be located and covered just by one tablespace.

3 Indexes

Query evaluation and retrieval is a complex staged process. Current database systems define the execution of the SQL statement by these phases – *parse, bind, execute,* and *fetch.* Before the processing itself, the system evaluates, whether the query has not been processed, yet. If we have prepared and stored the execution plan, the database optimizer instructs us to load it from the memory *Library cache* and use it. The task of the *parse* is to get the best possible access plan to obtain data, it is based on the indexes, mostly B+tree formats [23]. In the *bind* phase, variables are expanded into literals. Afterward, the *execution* phase is started by loading data blocks into the memory *Buffer cache.* In the *parse* and *execute* phase, our solution optimization can be identified. Loaded data blocks are parsed and particular data are extracted. They can be, however, non-actual reflecting the begin point of the SQL statement evaluation. If so, *Log buffer* comes into play. *The undo image* is built to the defined timepoint, respectively *SCN*. The *fetch* phase operates the result set composition and providing it to the client.

When dealing with the data tuples, the most important part of the evaluation and processing is based on selecting the access method. In principle, there are two options. If the suitable index is present in the system, based on the conditions of the statement, it will be used [21]. Regardless of the index type, it will be used to obtain data completely (if the whole result set can be composed of that – all attributes necessary for the processing is part of the index) or it will be used for the data located inside the database. On the leaf layer of the index, there is a pointer to the data called *ROWID* consisting of 10 bytes specifying data file, data block, and position inside the block, where the data row resists [3]. Thus, the evaluation is defined by two stages – index usage and data location, if necessary. The most significant aspects of the optimizer decision making are database object statistics and index suitability, which is expressed by the attributes forming it, as well as the order of attributes. As a consequence, mostly in the dynamic complex system, many times, the whole table has to be scanned (by *TAF* method). It means, that each block associated with the table is evaluated separately, sequentially block by block. From the physical perspective, each block has to be transferred to the memory *Buffer cache* and evaluated there. The granularity of the transfer is the block itself, which can be filled partially or even totally empty. As a result, the processing is too time and resource-demanding. Adding a new index does not solve the problem. If the queries are dynamic and vary significantly, it is not possible to define all possibilities. Moreover, it has additional demands on disc storage, memory, and performance, as well. To ensure the correctness of the result, it must be ensured, that the index is always up date, meaning, that any change on the data performed by the destructive *DML* (*Insert*, *Update*, *Delete*) must be applied to each index. Therefore, it is clear, that adding new indexes does not bring the desired power to the system. One of the solutions is based on autonomous index management based on the workload. Autoindexing [8, 13] is the technique for evaluating the total additional costs of adding and removing index based on the queries, as well as data changes weight. The disadvantage of this approach is the fact that it is a reactive form managing change, thus at first, a new impulse must be created for the reassessment of indexes, and only then the rebalancing itself can occur. This means that in the initial stages it is still necessary to use the *TAF* access method scanning all data blocks associated with the table sequentially. Our proposed solution aims to fill the gap and limit the necessity to use the *TAF* method.

In non-relational databases, the principles and contribution of indexes are a bit different. They often highlight the consistency optimizing for high throughput of the *write* operations. Therefore the results can be *stale* until the index incrementally rebuilds newly added documents [16, 29]. In *RavenDB*, data outputs do not need to reflect the up-to-date image, although it is possible to force the system to wait for non-stale results explicitly. Other systems, like *PostgreSQL* or *MongoDB*, are aware of consistency, which can have, unfortunately, a strong impact on performance [11]. In a relational database, the optimizer selects the best suitable method for obtaining data. In principle, it can be done either by using index or data blocks are scanned sequentially. In contrast to the *RavenDB*, all data queries must use an index interface to locate data, there is no other way. It offers two modes – *static indexes* and *auto index* mode, which generates index, whenever the requirement does not pass the existing static index [29]. In non-relational databases, various formats and structures can be identified, like explicitly written in C# or JavaScript (*RavenDB*), JSON (*MongoDB*), etc. [5, 15]. The rebuilding index is done in two phases – a new index is built and after the construction, the original structure is removed. Such a principle would not be suitable for the relational form if a strong data stream can be identified. Moreover, it is impossible to use the index, which does not reflect all changes. As evident, if the attribute holds a *NULL* value, a particular tuple is not stored in the relational index resulting in the necessity for sequential scanning. Thus, from the security point of view, all approved transaction changes must be immediately present in the relational index.

4 Removing the Impact of TAF

4.1 State of the Art

Our main contribution is to lower the usage of the *TAF* method to the minimum. If the block fragmentation is present, the problem is even deeper. The first solution was introduced in [20, 23]. It is based on the fact, that each index referencing all data tuples can be used as the data locator. In the leaf layer of the index, there is a pointer layer consisting of the *ROWIDs*. In this case, however, the system uses just the block identification extracted from the *ROWID*. Thus, the particular block was loaded into the memory based on the block identification and afterward, it was completely scanned to locate any data tuple inside. It provided a significant performance benefit [20, 21]. To summarize the impacts, we propose these results extended by the new technique defined in this paper. The solution defined in 2020 was based on the *Master index*, which was selected as the best suitable index holding all data references. It procedures either the total size of the index, its depth, as well as costs of the loading index into the memory. Thus, *Master index* selection was done always in a dynamic manner based on the current situation inside the system, whereas some index can be already present in the memory *Buffer cache*. In that case, in principle, it is better to use it, although the size is higher in comparison with other indexes, on the other hand, the loading process into the memory can be skipped – particular data are already there. Similarly, the index can be partially loaded into memory. All these characteristics are maintained by the *Master index* process as the extension of the database query optimizer [21].

The limitation of the above approach is based on the leaf layer. If the block consists of several rows, a particular block will be listed in the leaf layer multiple times, whereas block identification is just the part of the *ROWID*. In this section, we highlight two techniques for improving performance. First of all, when fragmentation of the block is present, it can many times happen, that the row is not in the block, we assume. *ROWID* on the leaf layer points to the row, where it should persist. It is, in principle, correct. However, just in the ideal conditions. If the multiple *Update* statements occur in one data row, many times, it consequences in shifting the row into another data block, whereas the original size is not suitable – row size is extended and the result data cannot fit the original position. The problem causes migrating rows into different data blocks [20]. Whereas there is no pointer from the physical data to the referencing indexes, it is not possible to mark the *ROWID* change in the indexes – it would be too demanding checking all developed indexes and evaluate the relevance of the change for it by traversing the index to the leaf layer [23]. Therefore, in the original block and position of the original row, the new pointer is added to another block, where the data are present. Figure 2 shows the principles. Notice, that the row can be migrated multiple times. In that case, therefore the block *A* has to be loaded based on the index *ROWID*. Unfortunately, the data themselves are not present there, there is just another *ROWID* pointer to the different block (*B*), which must be loaded and evaluated, too. Based on the model shown in Fig. 2, three blocks must be loaded before the data are located, which is not optimal. If the index pointed to the direct position of the row, it would be necessary to use just one I/O disc operation, in comparison with three ones. On the other hand, it is not useful to rebalance the index each time. Although this would increase data retrieval performance, the reconstruction of the index itself, and the cost of performing it would very shortly erase the expected improvement significantly [23].

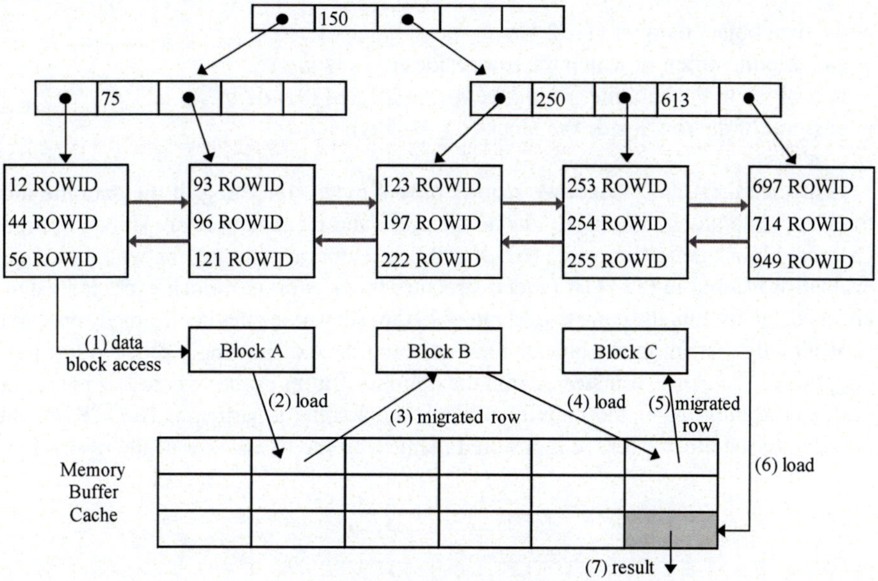

Fig. 2. Migrated row

The second problem is, that the index consists of the position of the row inside the block, however, such information does not bring the benefit for the data location without index condition management itself and the block identifier must be extracted for each value followed by the identification of the duplicate value. In current solutions [23], the block is accessed only once and processed completely. The task is to remove the impact of the full memory *Buffer cache* by limiting the transfer of the empty blocks from the database. Note, that this information cannot be obtained directly without block loading into the memory for the evaluation. If the workload is high, to load data into the memory, free space must be identified, thus many blocks from the memory are to be loaded to the database before [10]. Limitation of the number of blocks to be processed and evaluated can, therefore, for sure, improve performance.

4.2 Own Proposed Solution – Flower Index on Block Granularity

Analysis of the system performance and data access impacts is key for ensuring performance, by identifying too demanding parts of the processing. As evident, to ensure the performance by limiting TAF method usage, we need to restrict the *ROWID* values into the individual block referencing.

Our contribution aims in three developed methods. All of them are covered by the Flower index approach on the block granularity. The first solution (***MODEL 1***) is based on the original B+tree index, however, on the leaf layer, there is no *ROWID*, but just block identifiers (*BLOCKID*) is present. As a result, there is no necessity to extract block information from each *ROWID*. Moreover, the total size demands of the index are lowered, as well. The original *ROWID* value consists of the 10 bytes:

- the data object number (*1–32 bits*),
- data file identifier, in which the row resides (*33–44 bits*),
- data block in the data file, where the row is located (*45–64 bits*),
- position of the row inside the block (*65–80 bits*).

For the BLOCKID value, we require only 4 bytes consisting of the data file identifier and data block definition. The architecture and management are shown in Fig. 3. Extracted block definition value (BLOCKID) is temporary (during the whole statement evaluation) stored in the PGA (added structure block_storage). Each extracted value is checked against this structure to evaluate, whether such block has been already processed (it is already present in the block_storage) or not. If not, the whole block is located in the physical database, transferred into the memory Buffer cache, where it is parsed and evaluated totally. After the individual block processing, a particular BLOCKID value is added to the block_storage in a sorted manner. In the first stages of the research, the

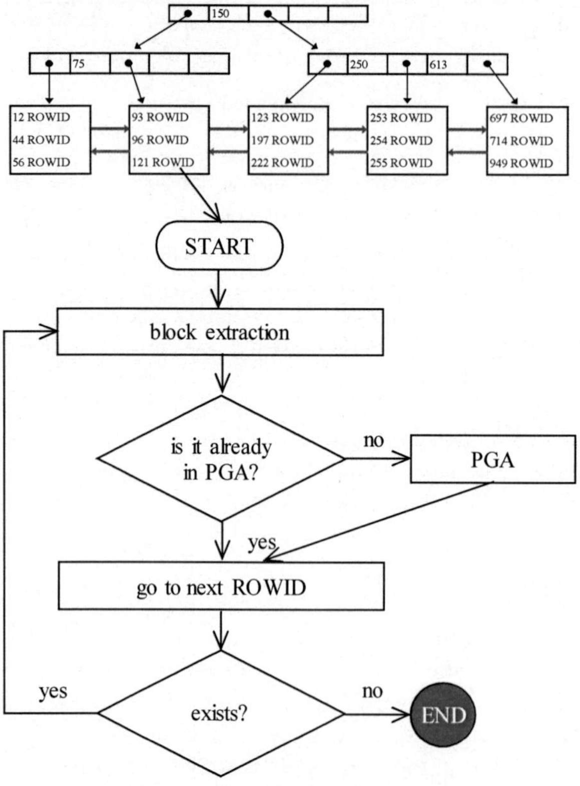

Fig. 3. BLOCKID management

block_storage structure was stored in the sorted linked list. Nowadays, a binary tree is used. The evaluation management and dataflow in the diagram are shown in Fig. 3.

The limitation of the proposed solution is the location of the block_storage structure. It is built for each statement and located in the PGA, invalidated after the query processing itself. Whereas it does not reflect any data change (new Insert, Update or Delete), it cannot be used multiple times, even in the same session reflecting the same table. Whereas the block_storage structure granularity is the whole table, if it reflected data changes, it would be possible to use it multiple times. Note, that after the processing itself, it consists of a list of all blocks consisting of at least one data tuple inside for the particular table. Therefore, in the second proposed solution (MODEL 2), we separate the block_storage structure and locate it in the shared memory (SGA) named shared_table_block_storage (STBS). In this case, it has two improved properties. It is sharable among the whole instance for each user and session referencing the same data object. Moreover, it is proof of any change, which is automatically incorporated into the structure.

4.3 Physical Implementation of the Shared_table_block_storage Structure

STBS structure is shared and located in the instance memory. In ideal conditions, it can always be located in the memory and there is no necessity to free the memory to serve another data block. If not, it is transferred to the temporary tablespace, from which it can be reallocated. Physical storage depends on the user selection provided by the parameter *store_block_definition* value, which can hold either value *True* or *False*. By default, it is not stored durable, thus the default value is *False*. In this case, if the instance is shut down, particular memory is freed and data are removed. After the instance is again available, the *STBS* structure is built during the first table reference and maintained automatically. Vice versa, if the value is set to *True*, before the instance *closing*, the memory structure *STBS* is *checkpointed* and copied into the physical database in the compressed format. During its instance opening, a particular structure is copied into the memory again and maintained there (original database space is then freed). The advantage of this approach is the fact that at the first request for the use of this structure it is already prepared in the memory. The data flow diagram of the approach is in Fig. 4.

The database structure allocation unit is not the block itself, but the set of blocks forming the *extent* covered by the position of the last associated block to the table – *High Water Mark (HWM)* position pointer [21]. Individual blocks inside the extent are interconnected, extents themselves are interconnected using the linked list, as well. The third proposed model (***MODEL 3***) uses its own index structure, however, there are no references to the rows, nor blocks. It is defined by the *B+tree linking extents*. These extents are managed by the added instance process *Extent_balancer*. This process aims to reallocate data across the extent – individual blocks are filled from the top, by executing *Update*, respectively *Delete* statement, *Extent_balancer* process ensures, that there is no gap between data – it cannot happen that there is an empty block followed by a block containing data. Thanks to that, if the database optimizer by loading process identifies an empty block, processing continues with the following extent, because it is ensured that there can be no more data in the given extent.

Improvement of the model 3 is delimited by the physical structure, in which the *Extent_balancer* does not move data physically to ensure sorting blocks based on the

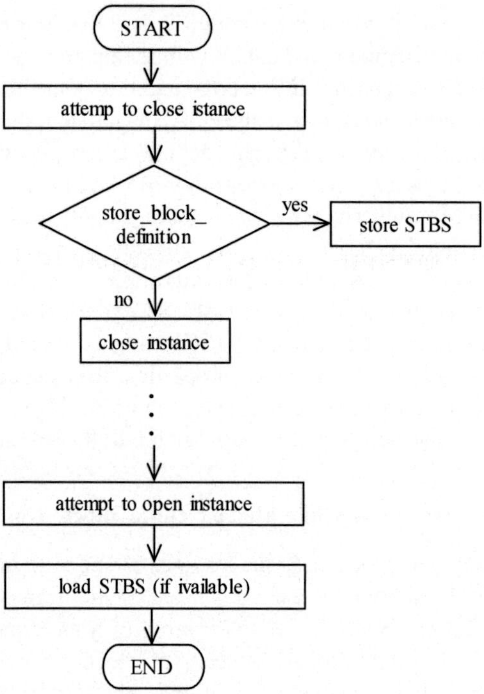

Fig. 4. Data flow – opening and closing instance process

amount of data, but rebalancing is done just by the shifting pointers to the individual blocks (***MODEL 4***). The structure of the extent is divided into six separate structures denoted by the flags specifying the amount of data inside. These properties are stored in the extent header:

- empty block - data amount 0%.
- data amount in the range of (0, 25)%,
- data amount in the range of <25, 50)%,
- data amount in the range of <50, 75)%,
- data amount in the range of <75, 100)%,
- full block – data amount 100%.

4.4 Rebalancer

Fragmentation is a complex problem of the current database system if the attribute data type size is not fixed. Undefined values cause changes in the size of the data record. As already described, in *model* 3, there is a rebalancing strategy to shift full blocks to the top for sooner evaluation and processing. If the value to be found is unique, it is also a higher probability to locate it in the fuller block than a block containing only one record. There is, however, another strategy dealing with the block itself. There can be a problem with the data fragmentation inside the block itself. Rebalancing and free space merging on the block granularity is not done automatically. First of all, it would require additional sources for each operation. Block occupation percentage is calculated and stored based on the definition, either in the data dictionary or on the block level [13, 14]. If it is clear, that the new row fits the block, only in that case, we structuralize and rebalance the block. It is more efficient to shift the data across the block in comparison with allocating a new block. It should be emphasized, that the rebalancing at the data block level causes migrated rows. The record is no longer in the original data position, but it is moved to another place within the same block. Thus, the *ROWID* value will not be accurate, but a pointer to another position in the given block will be present. The fact, that the whole block must always be loaded into the memory in its entire form ensures no performance reduction at all.

Migrated row from the defined index structures point of view causes a significant problem. Original *ROWID* value points to the block, where the data are not present, only another pointer is there to locate the different position. In the ideal condition, it points to the same block, but in another position. Mostly, however, it points to another block, which has to be loaded into the memory. This negative aspect applies just to the indexes, our proposed location models are prone to it. Therefore, as the extension of our solution, if the migrated row is present and processed by the index, a new position value replaces the original one inside the index to remove the impact of the migrated row. It is ensured by the added instance parameter *migration_manage* set to value *True*. If the value is set to *False*, in that case, migrated rows are still present, but the index does not reflect the change by replacing the pointing value. Such value (*False*) should be used in environments, where many data changes occur consequencing in shifting data in often rate. In that case, it would not bring you the benefit, whereas before new index usage, a particular data row would be moved and new migration is present.

5 JOIN – Linking Data Across the Tables

When tables are to be joined, the *join operator algorithm* must be selected to implement logical joins between two sets of data. The existing system scenario is based on available indexes, up-to-date statistics, and the number of estimated rows in each data set [7]. In this section, we highlight four existing approaches covering the most significant principles – *nested loop, merge join, hash match,* and *adaptive join* [8, 12]. The *nested loop* is the simplest operation, for each record from the outer input (smaller table), matching rows from the inner output are matched. It is done by using the index approach - *index seek* method. The complexity of the query is $O(N * log(M))$, where the N and M are cardinality estimation of the tables. *Merge join* is currently the most efficient way to join two large data sets, whereas both input operators are executed only once. It requires indexes on both tables and uses the property, that both of them are index sorted. The complexity costs of the processing are $O(M + N)$. *Hash match* is used, if the previous types cannot be used (table is not indexed based on the join key). It uses the *hash function* [12], which takes one or more values and converts them to a single symbolic value. A *Hash table* is a data structure that divides all rows into equal-size buckets, where each bucket is represented by a hash table [18]. The system chooses the smaller of the two inputs to serve as the build input and will be the one to build a hash table in the memory, if available.

Otherwise, it is stored in the physical disc with all consequences on performance. If the hash table is built, the system gets the data from the larger table, compares them to the hash table using the hash function, and return any matched rows. As long as the smaller table is very small, this algorithm can be very efficient. But if both tables are very large, this can be a very costly execution plan.

The last method is a combination of already mentioned types and does not rely so much on the statistics, but proposes multiple scenarios. The selection of the access path is defined by the estimated number of rows produced from the previous step. Thanks to that, it can swap between the *Hash join* and the *Nested loop* dynamically.

6 Table Joining – Mapping Index

All the above techniques rely on the existence of the unique index for the join key at least on one table. In dynamic systems, however, such a prerequisite may be broken. Generally, tables can be joined in any manner. One way or another, it is clear, that for the connection, every single record must be accessed, evaluated, and processed. The direct access can be replaced by the index checking, if suitable and available. If not, there is no other choice, the whole table must be scanned. Similarly, if the index is too old not reflecting the current situation – index nodes are never deallocated consequencing in the continual growth and cost growth. Our proposed access techniques described in Sect. 5 can be used for joining, as well. There is, however, one extra solution for joining the *primary key* values, respectively its candidate to the *foreign key* in a strict manner. In existing complex information systems, tables are usually large consisting of not only current valid data, but the whole evolution of the states over time, thus the *Nested loop* technique is not performance effective. If both tables have mechanisms for direct sorting, the index is used to control the connection. If so, the processing provides particular data pointers (*ROWID*), by which the whole row of the table, respectively requested attributes that can be loaded. Thus, the *index scan* method category is used. If the key for the foreign key is not present (as the result of the additional performance demands for destructive *DML – Insert*, *Update* and *Delete*), *Hash join* has to be used, mapping the whole table join key into buckets to the memory. If the sufficient memory size is not available, disc swapping is present resulting in poor performance. In the next paragraph, we propose our own solution characterized by the mapping index.

If the foreign key index is not present, it is clear, that the performance will suffer. Therefore, we define the composite two-table index mapping join key from the master table (table delimited by the join key as the primary key) to the secondary (foreign key join key). The main part consists of the *B+tree* index based on the primary key. On the leaf layer of the index, there is not the only pointer to the data, but another structure serving join key mapping. The name of the structure is *Mapper* and its aim is to provide pointers to the foreign key table data directly. At the main level, there are directives and references to the main table, the *Mapper* refers to the records from the second table, defined by the linked list of *ROWID_FK*. Figure 5 shows the architecture of the mapping index. Bold text with the gray background is the *Mapper* object. The interconnection can be done automatically, no existing method has to be used, whereas the join operation is directly implemented in the *mapper index*. Each relationship between the tables is delimited by the *type* (identifying, non-identifying), *cardinality* (1:1, 1:N), and *membership* (mandatory, optional). Relationship type does not influence the structure of the mapping index. Cardinality delimits the number of elements in the *Mapper* structure. It is protected by the *mapper_limit* parameter. If no limitation is used, the value "*–1*" is used. Optional membership is specified by the value *0* for the *mapper_limit*. In that case, a *new virtual index node* in the main structure is added expressing the non-connected values of the *Mapper*. Physically, it is positioned in the *top right of the B+tree*, so these records are processed at the end. A *virtual node* is optional if mandatory relationship type is used, it is not present, whereas it would be always empty.

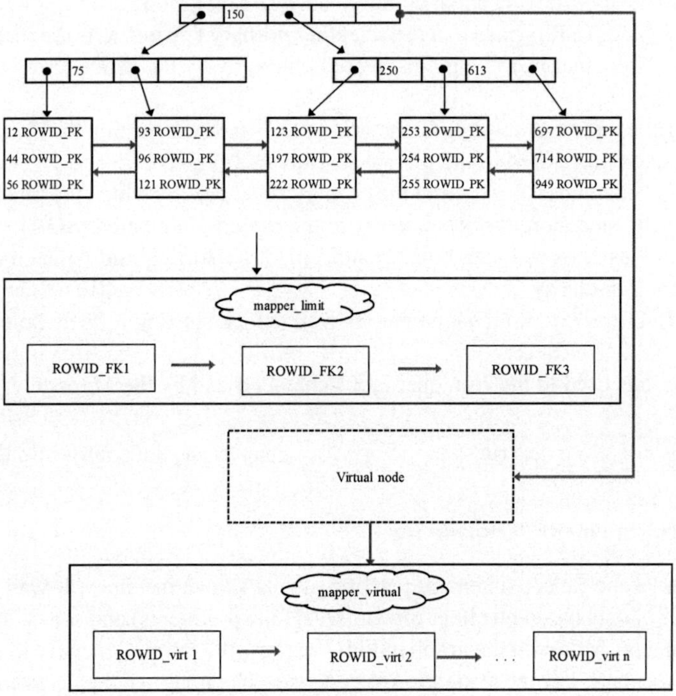

Fig. 5. Mapper architecture

7 Performance Evaluations

For the computational study and cost analysis, the relational platform of the database system *Oracle 19c Enterprise release 19.3.0.0.0 – 64 bit Production* was used. Computing server parameters are processor Intel Xeon E5620; 2.4 GHz (8 cores), 48 GB DDR 1333 MHz operation memory. For the evaluation, a table containing 10 attributes originated from the sensor environment was used, delimited by the composite primary key consisting of two attributes – sensor identifier and time measurement occurrence. The cardinality of the table was 1 million rows. The size of the individual rows was dynamic ranging from 98 bytes to 145 bytes, average size demands were 129 bytes. The second table has a similar structure, but the spatial extension is added. For the evaluation, no explicit index definition was used. The master index was always based on the primary key.

For the evaluation, these models were used. Each experiment was performed 10 times; the results show the average values:

- MODEL TAF highlights the existing architecture by scanning all data table blocks sequentially, whereas no suitable index is present.
- MODEL INDEX is a referential model providing the most optimal solution if the completely suitable index would be present. Such a model is used for evaluating benefits and additional costs in comparison with other models.
- MODEL MASTER is based on the selecting primary key index as the master for data locating. Note, that just one index created implicitly by the primary key is present in the table.
- MODEL 1 characteristics delimit the primary key index, by which the block identifiers are extracted and stored in the private session area (PGA).
- MODEL 2 characteristics delimit the primary key index, by which the block identifiers are extracted and stored independently in the shared global area (SGA).
- MODEL 3 extends the data block management by sorting and balancing blocks on the extent granularity.
- MODEL 4 uses extent management, as well. Block shifting reflects pointer consolidation.
- MODEL 5 is used in the Join operations characterized by the Mapper.

Models 1–5 were described in the previous chapters as the contribution.

7.1 Select Statement Performance

The results of the Select statement performance is shown in Fig. 6. It was specified to obtain 10% of data based on either time interval (5 experiments) and sensor specification (5 experiments). Values in the graph in Fig. 6 express the improvement of the processing time in percentage. For the evaluation, on one side The image of the performance specification can be delimited by two characteristics of the environment. On the one side, it is the TAF method, by which no index is used forcing the system to scan all blocks regardless of the fullness and fragmentation and on the other side, there is the specification of the most suitable index for the defined query. The proposed methods and models aim to get as close as possible to the index solution. The referential model for the processing is TAF reaching 100%. In the environment, 50% fragmentation is used. If there were not free space at all, processing time demands can be lowered up to the 70% value. The model index requires 10.9% of the processing time, which reflects 89.1% improvement (based on the assumption, that there is no data migration). The master index is used to locate data if a not suitable index is present. Whereas the condition is not based on the whole primary key definition (time range), the primary key index is not optimal. The processing required 17.3% of the processing time, which saved 82.7% of the processing time, in comparison with MODEL INDEX (reference 100%), there is 58.7% increased demands. MODEL 1 uses block definition composition in the PGA. It requires 17.3% of the processing time range. In comparison with MODEL INDEX, the slowdown is 39.4%. It is limited by the usage of non-shareable memory, the block identifiers are not updated and after the processing, the structure cannot be used repeatedly. MODEL 2 shifts the block definition management into the shareable memory section (SGA). On the one hand, it can be used by several sessions, moreover, it is not necessary to rebuild it every time. Besides, individual data changes are reflected in it. Reflecting on the reached result, the processing time requirement was lowered to 13.4%, which reflects the 86.5% improvement for TAF. In comparison with MODEL 1, there is 13.4% improvement, whereas it can be shared. If the structure is stored physically in the database (delimited by the *store_block_definition* parameter), performance demands are lowered to the value of 12.8%. Model 3 extends the management by the extent of consolidation sorting. There is a pointer in each extent, after which no data can be present. Thanks to that, these blocks do not need to be loaded and evaluated, at all. MODEL 3 reaches 12.5% of the total processing time demands of TAF. It reflects the improvement of 87.5% for TAF, in comparison with MODEL 2, there is improvement ranging to 7.2%. MODEL 4 uses consolidation specified by the pointer linking. In that case, processing time demands are 12.0%.

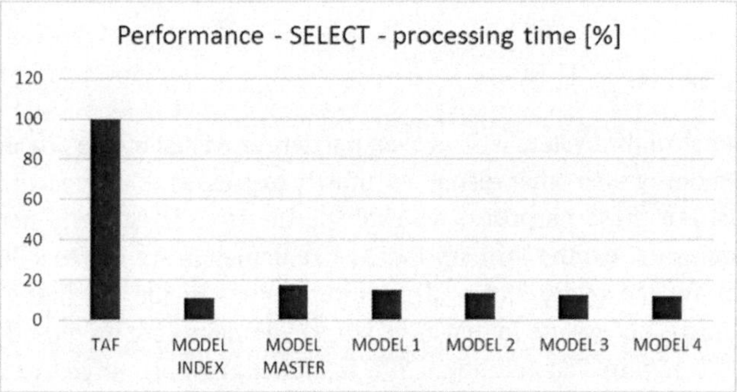

Fig. 6. Processing time performance – Select statement

7.2 Destructive DML Statement Performance

It is evident, that the *Select* statements benefit, if the index, respectively other effective solutions accessing data, is present. On the other hand, it is not effective to develop many indexes, whereas other *DML* statements are negatively influenced – all changes must be applied to all indexes before the operation itself. In this section, performance impacts are present. Each data object (*1000 objects are present*) was updated *1000 times.* Note, that invalidating object means, that it is removed from the table (*Delete*). Vice versa, if the value is later present, transforming the object into a valid state is represented by the *Insert* statement. In this experiment, the invalidation and validation rate is using *100.* In the graph, additional demands to the processing are expressed in percentage. Similarly, to the previous evaluation, experiments were performed 10 times, results in Fig. 7 show the average values. The TAF method does not use any index, so no additional demands are present. The optimal index definition increases the demands using 6.2%. The primary key definition is a bit lowered (4.3%), whereas it is used for ensuring integrity and data location for the update, as well. The additional index itself demands are 1.9%. MODEL 1 does not reflect changes performed, it is stored in PGA, therefore no additional demands are present. Other models have a very similar increase in processing time demands ranging from 5.9% (MODEL 2) to 6.2% (MODEL 4).

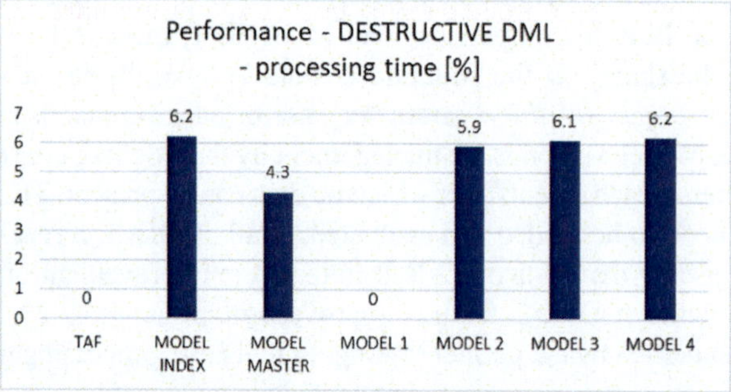

Fig. 7. Processing time performance – Destructive DML statements

7.3 Size Demands

The performance of the system is, however, not delimited just by the processing time. It is inevitable to monitor also other resources, mostly expresses as the costs of the additional size demands. For these purposes, we evaluate the storage capacity necessary for the objects. When using just the primary index, size demands are increased to 114.7%. An optimal index for the query stores all required values inside the index itself, thus no physical loading is necessary. In that case, storage demands are increased to 173.9%, the additional index requires an additional 59.2%, the original primary key index represents 14.7%. If the block index object is stored in the file system, 18.6% of additional capacity is required. Management on the extent granularity sorts just the data to shift data to the top-level blocks for sooner loading. As evident from Fig. 8, it does not have additional

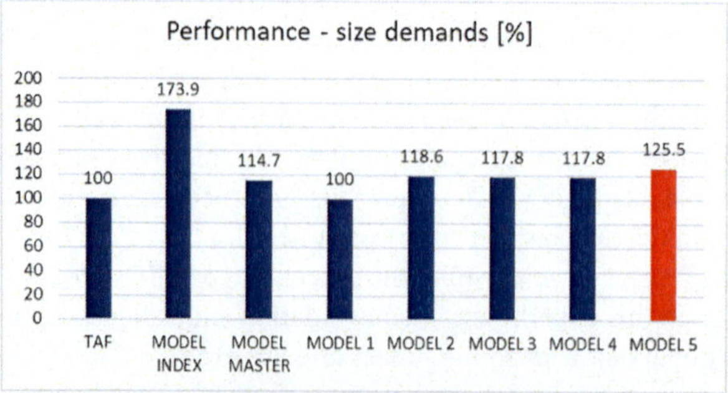

Fig. 8. Size demands

storage demands, respectively it is minimal improvement. Fig. 8 shows the results of the physical size demands monitoring.

In the above Fig. 8, there is also the specification of the size demands for the MODEL 5, which uses B+tree for the primary key index extended by the foreign key reference in the second table using the *Mapper*. The additional size demands are 25.5%, which expresses 73.5% increase in comparison with just one definition of the primary key. Note, that if two separate indexes were used, the demands would be 32.7%. Table 1 shows the results. Values represent the additional demands in percentage.

Table 1. Size demands for table join operators

Index	Size demands [%]
Primary key index	14.7
Separate indexes for the primary key and foreign key	32.7
Mapper	25.5

The size and structure of the database do not have a significant impact on the performance and proposed solutions are robust. Moreover, defined techniques can, in comparison with existing applications, benefit, if the data fragmentation across the data blocks is present.

7.4 Table Joining Performance

The relational paradigm is defined by the tables interconnected via individual relationships. If data need to be obtained from several tables, data must be joined together. Mapping rows can be done using various join methods described in Sect. 5. The best solution is provided by the *Merge join* if indexes on both sites (primary and foreign keys are present). If none of the indexes is present, data must be sorted in the first phase requiring a significant amount of processing time and resource consuming. The referential model is index provided requiring 40.4% of processing time in comparison with the TAF method (100%). Individual models for data projection and selection can be used in this case, as well. MODEL 1 requires 76.3%, whereas the block pointers are built dynamically during the execution. If the solution is shifted into the shared memory (MODEL 2) reflecting changes, processing requires 48.2%. In comparison with *Merge join* (100% reference), there are 8.2% increased demands caused by the necessity to remove duplicates from the block list in the memory. Extent management (MODEL 3 and MODEL 4) solves the problem only in a slight manner and does not provide relevant improvement. The results are in Fig. 9.

The proposed solution using the Mapper object provides the best performance. It is caused by the direct pointer to the other table references. There is no necessity to merge data and compare values of the PK and FK, all the interconnection is directly located in the index Mapper. On the main structure, a list of individual PK values is defined in

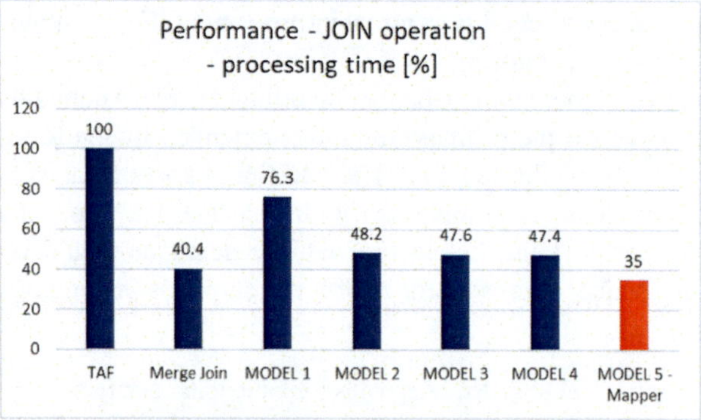

Fig. 9. Processing time performance – JOIN

the B+tree structure. FK references are present on the leaf layer for each element. It is managed automatically if any data change occurs.

8 Conclusions

Relational database systems are still powerful solutions for managing data in the information systems. Its main advantage is relational algebra support [30] and transaction covering ensuring reliability, consistency, and durability by passing all integrity rules. Nowadays, it is evident, that the number of data to be managed is significantly growing every year. Data evolve reflecting individual Update operations. Historical data with poor information value are removed, vice versa new data images are inserted dynamically. As a result, data fragmentation is present either in the data files, as well as indexes do not provide optimal performance due to many nodes, which are not deallocated. The aim of the database management from the user point of view is to ensure performance, get the data as soon as possible. If the index is present, a robust solution is provided. However, index management requires data change application for each destructive operation. Thus, it is not possible to develop all suitable indexes. If there is a huge number of different queries to be executed, many times, a particular table must be fully scanned using the Table Access Full method (TAF). This paper consists of the existing system analysis covered by the database system architecture – background processes and instance memory structures. Each time data are to be obtained, data blocks must be transferred into the memory for the processing if they are not present there yet. If the data fragmentation is present, the final performance is not relevant, whereas even empty blocks must be loaded. For data selection and projection, several new models are proposed. Unlike the standard index approaches, the proposed solutions use block locator instead of row locators. Thanks to that, the whole block is processed after the loading. Solution objects can be located either in the private area of the server session process, however, if the model is shared among the whole instance and changes are reflected, significant performance improvements can be reached.

In the final part, principles of data joining across the data tables are defined as proposing the Mapper between master and child table. The core part is formed by the B+tree index supervised by the reference list in the leaf layer.

Future research in this field will be oriented to the block granularity itself and data distribution across the tablespace partitions. Thanks to that, the size of the block can vary based on the data. Moreover, each partition can be optimized separately by defining storage parameters, structures, and index approaches, as well.

Acknowledgment. This publication was realized with support of the Operational Programme Integrated Infrastructure in frame of the project: Intelligent systems for UAV real-time operation and data processing, code ITMS2014+: 313011V422 and co-financed by the European Regional Development Found.

References

1. Abdalla, H.I.: A synchronized design technique for efficient data distribution. Comput. Hum. Behav. **30**, 427–435 (2014)
2. Abdalla, H.I., Amer, A.A.: Dynamic horizontal fragmentation, replication and allocation model in DDBSs. In: International Conference on Information Technology and e-Services, ICITeS 2012, Tunisia (2012)
3. Alghamdi, N., Zhang, L., Zhang, H., Rundensteiner, E., Eltabakh, M.: ChainLink: indexing big time series data for long subsequence matching. In: IEEE 36th International Conference on Data Engineering (ICDE) (2020)
4. Amin, M., Rahman, R.: Universal database access layer to facilitate query. In: Ninth International Conference on Digital Information Management, ICDIM 2014 (2014)
5. Aydin, B., Akkineni, V., Angryk, R.A.: Modeling and indexing spatiotemporal trajectory data in non-relational databases. In: Managing Big Data in Cloud Computing Environments (2016)
6. Bai, Y., Bhalla, S.: Introduction to databases. In: SQL Server Database Programming with Visual Basic.NET: Concepts, Designs and Implementations (2020)
7. Bottoni, P., Ceriani, M.: Using blocks to get more blocks: exploring linked data through integration of queries and result sets in block programming. In: 2015 IEEE Blocks and Beyond Workshop (2015)
8. Bryla, B.: Oracle Database 12c The Complete Reference. Oracle Press (2013). ISBN - 978-0071801751
9. Bulysheva, L., Bulyshev, A., Kataev, M.: Visual database design: indexing methods. In: 2018 Sixth International Conference on Enterprise Systems (ES) (2018)
10. Burleson, D.K.: Oracle High-Performance SQL Tuning. Oracle Press (2001). ISBN - 9780072190588
11. Chopade, R., Pachghare, V.: MongoDB indexing for performance improvement. In: Advances in Intelligent Systems and Computing, vol. 1077 (2020)
12. Dan, T., Luo, C., Li, Y., Zheng, B., Li, G.: Spatial temporal trajectory similarity join. Lecture Notes in Computer Science, LNCS, vol. 11642, pp. 251–259 (2019)
13. Date, C.J.: SQL and Relational Theory - How to Write Accurate SQL Code. O'Reilly Media (2015). ISBN - 13:978-1449328016. ISBN - 10:1449328016
14. Delplanque, J., Etien, A., Anquetil, N., Auverlot, O.: Relational database schema evolution: an industrial case study. In: IEEE International Conference on Software Maintenance and Evolution, ICSME 2018, Spain, pp. 635–644 (2018)
15. Drzymala, P., Welfle, H.: Support JSON standard for storing and processing data in the Oracle environment. Przeglad Elektrotechniczny **96**(2) (2020)
16. Eini, O.: The pain of implementing LINQ providers. Queue **9**(7) (2011)

17. Feng, J., Guoliang, L., Wang, J.: Finding top-k answers in keyword search over relational databases using tuple units. IEEE Trans. Knowl. Data Eng. **23**(12), 1781–1794 (2011)
18. Ivanova, E., Sokolinsky, L.B.: Join decomposition based on fragmented column indices. Lobachevskii J. Math. **37**(3), 255–260 (2016)
19. Kvet, M., Kršák, E., Matiaško, K.: Temporal database architecture enhancements. In: Proceedings 22nd Conference of Open Innovations Association FRUCT, pp. 121–130. FRUCT Oy, [S.l.]. ISBN 978-952-68653-5-5
20. Kvet, M., Matiaško, K.: Analysis of temporal data management in the intelligent transport system. In: DISA 2018: IEEE World Symposium on Digital Intelligence for Systems and Machines: Proceedings, pp. 151–157. Institute of Electrical and Electronics Engineers, Danver. ISBN 978-1-5386-5101-8
21. Kvet, M., Matiaško, K.: Temporal flower index eliminating impact of high water mark. In: Innovations for Community Services: Proceedings, 1 Vyd., pp. 85–98. Springer International Publishing AG, Cham (2018). ISBN 978-3-319-93407-5
22. Lan, L., Shi, R., Wang, B., Zhang, L., Shi, J.: A lightweight time series main-memory database for IoT real-time services. Lecture Notes in Computer Science, LNCS, vol. 11894, pp. 220–236 (2020)
23. Lin, H., Chen, Ch.: Using compressed B+-trees for line-based database indexes. In: 2006 IEEE International Symposium on Signal Processing and Information Technology (2006)
24. Maran, M., Paniavin, N., Poliushkin, I.: Alternative approaches to data storing and processing. In: International Conference on Information Technologies in Engineering Education (Inforino) (2020)
25. Maté, J.: Transformation of relational databases to transaction-time temporal databases. In: Engineering of Computer Based Systems (ECBS-EERC), 2nd Eastern European Regional Conference (2011). ISBN - 9780769544182
26. Ochs, A.R., et al.: Databases to efficiently manage medium sized, low velocity, multidimensional data in tissue engineering. J. Vis. Exp. JoVE **153** (2019)
27. Padmanabhan, S., Malkemus, T., Jhingran, A., Agarwal, R.: Block oriented processing of relational database operations in modern computer architectures. In: Proceedings 17th International Conference on Data Engineering (2001)
28. Smolinski, M.: Impact of storage space configuration on transaction processing performance for relational database in PostgreSQL. In: 14th International Conference on Beyond Databases, Architectures and Structures, BDAS (2018)
29. Song J., et al.: Haery: a Hadoop based query system on accumulative and high-dimensional data model for big data. IEEE Trans. Knowl. Data Eng. **32**(7), 1362–1377 (2020)
30. Vinayakumar, R. Soman, K., Menon, P.: DB-learn: studying relational algebra concepts by snapping blocks. In: International Conference on Computing, Communication and Networking Technologies, ICCCNT 2018, India (2018)
31. Zhang, W., Ross, K.: Exploiting data skew for improved query performance. IEEE Trans. Knowl. Data Eng. (2020)

Author Index

J. Paralič et al. (Eds.): DISA 2020, AISC 1281, pp. 203–204, 2021.
https://doi.org/10.1007/978-3-030-63872-6